D0820934

POWER STRUGGLE

POWER STRUGGLE

The Hundred-Year War over Electricity

RICHARD RUDOLPH
AND
SCOTT RIDLEY

1817
HARPER & ROW, PUBLISHERS, New York
Cambridge, Philadelphia, San Francisco, Washington
London, Mexico City, São Paulo, Singapore, Sydney

 For information address Harper & Row, Publishers, Inc., 10 East 53rd Street, New York, N.Y. 10022. Published simultaneously in Canada by Fitzhenry & Whiteside Limited, Toronto.

FIRST EDITION

Designer: Jénine Holmes

Copy editor: Ann Finlayson

Index by Auralie Logan

Library of Congress Cataloging-in-Publication Data

Rudolph, Richard.
Power struggle.

Bibliography: p.
Includes index.
1. Electric utilities—Government policy—United States—History. I. Ridley, Scott. II. Title.
HD9685.U5R73 1986 363.6'2'0973 85-45227
ISBN 0-06-015584-1

---

86 87 88 89 90 RRD 10 9 8 7 6 5 4 3 2 1

This book is dedicated to
Graham Ridley
and
Rick, Robin, and Suzanne Rudolph
and to
all those who have labored to promote democratic control
over one of the world's most powerful industries

Contents

Acknowledgments

There are many people to thank for their support, assistance, and comments on the material contained in this book. They include Eileen Austen, Robert Backus, Stephane Baile, Jeff Brummer, Katherine Durso, Anna Gyorgy, Sally Hindman, Ronald McKensie, Alan Nogee, Vic Reinemer, Kirk Stone, Donald Worster, and the many individuals who gave their time for interviews.

We are also indebted to our copy editor, Ann Finlayson, for her careful checking, to Aaron Asher for his insight and advice, and to Gloria Stern for her guidance.

The staff of many organizations and agencies also provided thoughtful assistance for this work. Among them: American Public Power Association, Department of Energy, Edison Electric Institute, Energy Information Administration, Environmental Action Energy Project, Environmental Task Force, Library of Congress, National Archives, National Association of Regulatory Utility Commissioners, National Rural Electric Cooperative Association, North American Electric Reliability Council, Safe Energy Communications Council, Securities and Exchange Commission, Center for Renewable Resources, and University of Massachusetts/Boston.

Finally, we'd like to thank the Fund for Investigative Journalism, which provided a grant for work that became the seed for this book.

No subject has come before you, nor will any come, that holds within it so vital and far-reaching a luxury as this over the daily life of the present and future men, women and children. . . . For good or evil, for economic freedom or industrial bondage—this change is upon us. What it shall bring depends upon ourselves, of a truth we are in the valley of decision.

Gifford Pinchot, Governor of Pennsylvania, on halting the growing influence of the power industry in 1925

Introduction

The electric power industry is the most money-intensive, pervasive, and politicized business in modern America. Its course over the next decade will decide a central part of the future of the nation. That course is now racked with turbulence and uncertainty.

The common perception is that the trouble with the electric industry appeared out of nowhere in the wake of the oil crisis of 1973 or the Northeast blackout of 1965, or with the beginning of nuclear power. But it goes back much further than that. Starting in the late 1880s, there emerged a struggle over who would control the resource of electricity, the geographic territories that went with it, and the increasingly central place it would occupy in the nation's economy. It was a bitter fight that raged through hundreds of cities and towns.

In the corporate paradise of nineteenth-century America, the question that erupted over electricity was whether it would be a "service" like water, guided by social policies, or a magical "commodity" whose growth and use would depend on corporate strategies. In other industrialized nations such as Great Britain, Canada, France, and Sweden, electricity would ultimately become a publicly owned resource. That was not to happen in the United States.

Out of the struggle between private and public interests, private companies came to dominate the industry and the politics surrounding it. Tied largely to expansion efforts, the turbulence came in waves in the 1890s, the 1920s, the 1930s, the early 1950s, and over the last decade. Amid ruthless battling against public interests, carefully calculated myths were put in place. History was popularly rewritten—for example, to exaggerate Thomas Edison's contributions and make him the father of the electric light and the benign figurehead for privately owned power companies. Behind the name of Edison or "public service," private power companies grew to be the massive "dividend machines" of Wall Street and helped fuel speculation that led to the 1929 stock market crash. Technology was also affected. Transmission networks that would have reduced consumer costs went unbuilt for decades because of private power's opposition. Nuclear power was rushed into commercialization in order to assure private companies' continued control of the electric industry and United States dominance of the international reactor market. And regulation of the industry was steadily undermined by a coordinated system of political influence and lobbying.

After decades of propaganda and political manipulation, our understanding and use of electricity has become more a function of that power struggle than a practical application of energy. Behind the message to "Live Better Electrically," consumption of electricity has been grossly inflated and technology used in a grand scam, which has depleted resources, degraded the environment, and fleeced consumers. Some energy experts have observed that if recommended conservation measures had been practiced, the nation's energy crises would never have occurred.

This book is an attempt to trace the intrigues tied to the century-long power struggle. Our purpose is to unearth the rich history of how this conflict between public and private interests has shaped and will shape the political future of the nation—our industrialization, the environment, and the concentration of economic power. In the words of power industry executives, community activists, consumers, environmentalists, Wall Street analysts, and government officials, the book tells the story of the evolution of a key industrial empire. And in many ways it shows how the conflicts over electric power and the crises we face have come full circle.

Sixty years ago, when Pennsylvania Governor Gifford Pinchot warned of the extensive social impact of an electric power empire, a few influential financiers controlled expansive electric holding company conglomerates that owned most of the electricity generated in the United States. The same electric holding companies also owned real estate businesses, railroads, streetcar lines, ice companies, water companies, financial firms, and far-flung operations in South America, Eastern Europe, and the Philippines. They exercised strong political influence at the local level, and behind closed doors many were directed by Wall Street speculators. Some of the most bitterly fought legislation of the New Deal era broke up the majority of those conglomerates.

Today, a more loosely knit electric power empire is once again attempting to expand its influence and operations in preparation for a new wave of centralized electrical expansion tied to visions of all-electric transportation, all-electric cities, and high-tech reindustrialization, which industry leaders say will require more nuclear and coal plants and the associated hazards of radioactive waste storage, the possibility of nuclear meltdowns, strip-mining, and acid rain. The public has been given no say over this corporate vision of the future, although each community is expected to quietly absorb the increasing social, economic, and environmental risks.

We're now at the early stages of this expansion. Witnessing the soaring bills and an environmental onslaught, people at the local level have been forced to take dramatic action. From Port Angeles, Washington, to towns like Moscow, Ohio, New Orleans, Louisiana, Brumley Gap, Virginia, and Concord, New Hampshire, they have crowded into town meeting halls, living rooms, and school auditoriums, and staged decade-long opposition to plans of runaway construction, environmental threats, and the erosion of local democratic control. This has been the latest wave of turbulence in a century of trouble. Now in the midst of a startling stall in the growth of the industry in the mid 1980s, it's not difficult to see how those surface struggles of the past decade will intensify as the nation moves more deeply into a high-tech and service-oriented economy and the pressure for electric expansion grows.

Already the federal Department of Energy has recommended that power companies spend more than a trillion dollars over the next twenty years on new coal and nuclear plants, tripling the industry's assets. And despite opposition in Congress, there is growing support to roll back the New Deal laws and allow power companies once again to form new holding companies and expand into operations such as real estate, telecommunications, industrial development, finance agencies, and countless other businesses.

Some energy experts hope that an "energy transition," using energy efficiency and solar power, will resolve the crises. But the use and development of energy-efficient and solar technology too is tied to political forces. At bottom, history shows that, rather than technology, politics will be the determining factor for where we will find ourselves in the coming decades. It also indicates that buried in the upheavals of a century-long power struggle and the dramatic confrontations we're now witnessing are the outlines of a deeper clash yet to come.

1. Warning Signs

> We're well aware that we're taking on the greedy contractors, the greedy bankers, the nuclear industry, and all the entrenched power management with their $125,000 salaries. But we can win!
>
> *Dorothy Lindsey, homemaker*
> *Hoquiam, Washington*

On a Sunday afternoon in February 1982, Dorothy Lindsey stood in the dense murmur under the caged lights and basketball nets of the Hoquiam High School gym watching as more than 3,000 residents of Grays Harbor County crowded inside. Two weeks earlier, she and a dozen of her neighbors had confronted the county utility commissioners over soaring bills for five nuclear plants. Costs for electricity had risen over 600 percent in five years and some families in the county were facing the loss of their homes, farms, or businesses. The commissioners told Lindsey and her friends nothing could be done. But they refused to believe it and began to organize. Two of the commissioners now sat behind Lindsey as she stepped nervously to the microphone. "For years Washington residents have sat back and let others run their public power," she said, letting her voice ring out over the sound system. "Those days are over!" The crowd stomped and whistled and cheered.

Twenty miles inland on Route 12, the cooling towers of two partially completed nuclear plants rose out of the steep wooded hills above the Satsop River. Another two hundred miles inland over the Cascade mountain range, three others were under construction in the rocky desert near Richland. Together, the five plants had made up one of the most ambitious nuclear projects in the world, aimed at helping fulfill a dream of boundless electric energy for the region. It was the same dream being promoted across the country as the road to the nation's "energy independence" and economic prosperity.

The meeting was a spark that would touch off a deepening conflict in the Pacific Northwest. It would bring the massive project tumbling down, shake Wall Street's bond market, and send a series of frantic proposals to Congress. Some observers viewed it as the beginning of the end for nuclear power. But nuclear proponents clung tightly to their vision of the future.

Four years later, radioactive winds from the tragic Chernobyl meltdown in the Soviet Union brought a spreading wave of concern over nuclear power. In North Carolina, Massachusetts, New Hampshire, New York, and other areas where reactors were under construction, local governments voted to withdraw from emergency evacuation planning in order to keep the plants from opening. Protests were also alive in more than a dozen states over the siting of radioactive waste dumps.

But while this opposition was building, federal officials were planning a "second coming" of nuclear power for the 1990s. Energy Secretary John Herrington and Nuclear Regulatory Commissioner Lando Zech, Jr., believed that the problem with nuclear power was not so much one of cost or safety but of public perception. They predicted a new energy crisis—an electricity shortage—unless construction of nuclear plants was restarted. Industry leaders raised a threat of the "lights going out in the 1990s" and urged formation of a special presidential commission. In Congress they pressed for passage of legislation limiting citizens' ability to intervene and clearing away institutional barriers to new nuclear construction.

Their efforts promised intensified turbulence. More than just a debate over nuclear power, it is a fight about visions of an increasingly electrified future with great environmental and economic impacts. At bottom it's an issue of democratic control and a conflict over who will decide the far-reaching risks and tradeoffs communities will face. Some federal officials see the quietly building crisis over electricity and its outcome as the key to the future of the United States. Perhaps more than any other event, the fiery series of protests that erupted in the Pacific Northwest reveal the politics at the heart of the growing debate.

At a small table behind Dorothy Lindsey, the two county commissioners, Arne Holm and Jack Welch, sat stunned by the thousands of angry people and the whistling and cheering that filled the Hoquiam High School gym. It had been a long hard road that brought them to this point. Within the dream of "boundless energy," as within any dream, were the seeds of its own undoing.

Planning for the plants had begun in 1968 as part of a program to build forty coal plants and twenty nuclear plants and make the Washington Public Power Supply System (WPPSS) in Richland the "energy czar" of the Northwest. Under the leadership of the WPPSS consortium, 118 public and rural cooperative power systems and three private companies agreed to finance the first three nuclear plants. Eighty-eight public and cooperative systems pledged their future rates to build two more.

Work did not progress smoothly. At Satsop, ground was broken in the fall of 1976, the rainiest time of year in one of the wettest places on earth. Half the cleared site went sliding downhill in huge mud streams. Contractors responded by covering fifteen and a half acres with plastic. As work schedules slipped steadily behind, claims emerged that contractors were milking the project for all it was worth. Reports also revealed that contractors jockeyed for position with their equipment while sometimes waiting weeks for material to arrive. Crews arrived to do work eighty feet up only to find that scaffolding had been taken down by other workers busy with another job. And welders complained that pipe hangers, which bracket miles of piping to the walls of the reactor building, had to be put up, taken down, put up, and taken down again. As the stories of boondoggles at the construction sites continued to surface, the acronym for the project was popularly changed from WPPSS to "Whoops." Through the late 1970s, soaring construction costs and chaos at the plant sites sent estimates for the project from $4 billion to over $15 billion. By 1982 the cost estimate for finishing the five plants rose to $24 billion.

Electric rates doubled and quadrupled. In response, consumers cut back on their electric use. But insulating walls and turning off lights seemed to do little to stop the soaring bills. In December 1981, with cost estimates still rising, project manager Robert Ferguson and the WPPSS directors looked at the decline in the need for electricity and were forced to admit at least two of the plants weren't needed. In January news broke that one plant at Satsop and one at Richland would be canceled. The total debt for funds already spent on the two unfinished reactors was more than $7 billion. In more than a hundred towns across Washington, Oregon, Idaho, and Montana, people were expected to pay off the bill in installments to bondholders over the next thirty years.

In Grays Harbor County, the fishing and lumber industries were in a slump, and electric bills had risen 600 percent. News of the plant cancellations was followed by the announcement of a new 50 percent rate increase, and more to come. Things had reached the breaking point. For some of the 34,000 families in the county, additional rate increases meant electric bills higher than mortgage payments and the possible loss of their homes or farms or businesses. Outraged by the prospect of thirty years of bills for electricity they would never receive, they responded en masse when they heard about Lindsey's meeting planned for Hoquiam High School.

Lindsey, a tall, attractive woman in her early fifties, proposed that the rate increases from the canceled plants be rescinded or the commissioners resign from office. Amid prolonged cheering she also proposed that the people of Grays Harbor County join with other consumers for a night of protest in which they would black out the state.

Although their demands were simple and straightforward, people knew that they were up against something larger than just the local commissioners. Several months before, an effort to establish voter control over the runaway costs of the project had been met with a $1.3 million countercampaign funded by major contractors, Wall Street brokerage firms, and large industrial companies in the region. Some of the companies, such as Merrill Lynch, claimed to be making the largest profits in their history on the project. Lindsey went at this point head on.

"We're well aware that we're taking on the greedy contractors, the greedy bankers, the nuclear industry, and all of the entrenched power management with their $125,000 salaries," she told the crowd. "But we can win!"

When the commissioners rose to speak, they were nearly drowned out by booing and catcalls. For the next two hours, the meeting became a public official's nightmare as housewives, senior citizens, ranchers, and loggers came forward to the audience microphones refusing to pay for the canceled plants and demanding that Wall Street "eat the debt." One state official who believed that the region's economic future was pinned on the massive nuclear project called the meeting a St. Valentine's Day massacre.

In the week that followed, news of the fiery gathering touched off a consumer rebellion that ripped across other WPPSS communities, uprooting the political establishment and throwing power companies into chaos. On February 21, three hundred consumers confronted power officials in Mason County. A day later 1,400 people gathered at the Port Angeles High School in Clallam County. Two hundred miles inland, residents of Ellensburg packed Morgan Junior High School. The spontaneous wave of anger then spread through Clark County in the south, Lewis County, Thurman County, north of Seattle to Snohomish, and over the state borders into Oregon and Idaho. Ranchers and housewives in Springfield, Oregon, staged a candlelight march and burned their bills. And throughout the region, consumers formed chapters of Lindsey's Irate Ratepayers organization, or similar groups. In response, Wall Street executives threatened trouble for financing of schools, bridges, and other local projects if the debt wasn't paid. But across three time zones, the disaster posed by paying the debt held a weightier impact.

In the Umatilla region of eastern Oregon, farmers dependent on electricity for irrigation looked at the local debt and estimated that their costs would increase $30 per acre, an extra $15,000 for a small 500-acre farm. On the rocky hillside near his orchard in central Washington, Stan Mettler observed that farmers would have to borrow more from the banks in order to make ends meet, or sell out. In the tiny town of Drain, Oregon, 600 residents were facing a bill of $4.5 million. Of the four lumber mills in town, two had already shut down, and the others threatened to move out of the area if electric costs climbed any higher. The same threat emerged from the aluminum industry. Jim Lazar, an economist from Olympia, estimated that payment of the debt would cause the loss of 100,000 jobs across Washington, Oregon, Idaho, and Montana.

As the national news media picked up the story of the showdown in the Northwest, it sparked similar but unrelated protests in California, Oklahoma, and Tennessee, where consumers were also facing bills for high costs of nuclear plants. Dorothy Lindsey said that for months after the meeting in Hoquiam her telephone never stopped ringing. Consumers called from New York and Texas, asking what they could do to oppose their soaring electric bills, and television crews from California and New York camped in her living room.

Behind the scenes, allegations emerged that the aluminum industry and other large power users had encouraged building the five reactors at once. With their twenty-year power contracts running out, the region's metals producers were interested in creating a supply of cheap surplus power. According to old press reports, they had been urging the towns to "build ahead of their needs" since 1975. At that time there had been a burst of controversy over the need for the plants. Many towns had been reluctant to become involved. At hearings on plans for the project, citizen and environmental groups had presented alternative power plans and argued for conservation programs. The regional power planning council was dominated by industrial interests, however, and disregarded the predictions of disaster if the plants were built. Donald Hodel, who was destined to become Secretary of Energy and Secretary of the Interior in the Reagan Administration, was head of the federal Bonneville Power Administration at the time. He supported the industrial interests and called the project's critics "prophets of shortage" and warned of brownouts, blackouts, and factory closings if the region's towns did not sign contracts to finance the project.

Five years later, in the wake of the project's collapse, Hodel tried to disclaim any responsibility. Others tried to blame inflation and mismanagement. But it was clear that in addition to obvious mismanagement, greed, and inflation, influential leaders had orchestrated the building of more plants than were needed and tried to keep them under construction long after it was reasonable to do so. Even after the first cancellations, when it became apparent that none of the plants was needed, area leaders met to stop the citizen opposition that had swept the region.

In July 1982, five months after the Hoquiam meeting, while the citizen groups were organizing ballot box campaigns for the fall election, twenty-one regional business leaders from financial firms, aluminum companies, and the nuclear industry met to plot their course of action. A nine-page memo was leaked from the "strategy breakfast" meeting. It revealed a plan to use a network of the local chambers of commerce, Northwest Chambers on Energy, to work with local media and threaten financial repercussions for the town's ability to finance schools, roads, and sewage lines if citizens refused to pay the debt on the canceled plants. Further, these business leaders wanted to use a new regional council—the Northwest Power Planning Council—to stress the need for the remaining three plants.

Robert Graham, president of the Seattle Chamber of Commerce, is noted as saying: "The Council is the most important creation of the new Act. Now it has to scare people. . . ."

Gordon Culp, an attorney for WPPSS, responded: "It is a mistake to focus on the council or plan. . . . It will be a backdrop against which we must prove something. . . . The campaign must emphasize the point: 'Do you want to throw away money you've already spent?' "

Eric Redman, counsel for the aluminum producers, added, "And pay a lot more for anything you want to buy in the future."

The group concluded that it needed a position paper to take to the editorial boards of local newspapers. Sam Voltenpest, a representative of the Tri-Cities Nuclear/Industrial Council, said: "You need it immediately. Then take it to other chambers—fan out. Other chambers can carry the message to local businesses; local business must carry it to local papers and employees."

Although unaware of the meeting and the plan, citizens were suspicious of the actions being taken by the industry. Some pressed for an investigation. Jim Duree, an attorney from Aberdeen, Washington, said that if a grand jury was convened in Tacoma, he believed indictments would be handed down against local power officials, financial advisers, and construction contractors. But no grand jury investigation was forthcoming. Caught between back-room maneuvering by industrial and financial leaders and the adamant citizen opposition, a bureaucratic paralysis settled over government leaders and the situation continued to worsen. As the estimates on future power needs dropped further, a third reactor was abandoned. Bailout efforts in the Washington state legislature and in Congress failed. Then, with money running out, work stopped at a fourth reactor. In the 1983 fall election, citizens won eighteen of twenty-four commissioner seats. But while this event held promise for conservation and alternative power in the long term, it was too late to stop the disaster that had mounted over the previous decade.

Finally on July 24, 1983, eighteen months after the fiery meeting in Grays Harbor County, the adamant refusal to pay for the unfinished plants, backed by court decisions invalidating the contracts between WPPSS and the power systems, rocked Wall Street with the largest municipal default in the nation's history. Chemical Bank of New York responded by organizing 75,000 bondholders and filing suits against 600 public officials for fraud. At Satsop and Richland, equipment was weatherized and the unfinished construction mothballed in hopes that work could be restarted at some point in the future. But the morass of suits was expected to take at least five years to work its way through the courts, and even if the debt was settled back on the region, people said they would still refuse to pay.

□ □ □

These events were only the beginning of a larger collapse. The dream of massive new electrification that had seized the Northwest was also tumbling into nightmare in other regions. Shock waves went through power company board rooms across the country as the specter of the financial debacle in the Northwest took shape. Nationwide, overbuilding of power projects had created an electricity surplus of nearly 40 percent in 1982. Some of these plants, such as Shoreham on Long Island and Seabrook in New Hampshire, were even more expensive than the WPPSS reactors. By early 1984 half the forty-eight nuclear plants under construction were in danger of being abandoned because of soaring costs and a lack of need for their power. January became known as the industry's blackest month. Day after day, newspaper headlines revealed a string of serious construction flaws and rising cost estimates which threatened work stoppages at plant sites in central and southern states.

Viewing the crippling wave of events, New York energy economist Charles Komanoff warned that the final "damage bill" for cost overruns and abandonment of nuclear plants across the country would amount to $100 billion. Other estimates put the damage at more than $200 billion.

Alan Nogee, a utility analyst for the Environmental Action Foundation, authored a comprehensive report in October 1984 which showed the "rateshock" of nuclear plants—the sudden increase in electric rates consumers have to pay above the cost of electricity from coal or oil-fired plants—to total $191 billion. The impact in a number of areas threatened to be just as severe as in the Northwest. Consumers in New Hampshire, for example, faced an annual debt on the Seabrook nuclear project that would triple electric rates and dwarf the state's annual budget. On Long Island, the Shoreham nuclear plant threatened increases that could eliminate up to 49,000 jobs, increase foreclosures and abandonments by as many as 2,000 per year, and push 11,000 families below the poverty line. The city of New Orleans faced rate increases of 70 percent, which would be shifted to consumers in other states, from the Grand Gulf nuclear plant. In Indiana, the difficulty was over the Marble Hill plants, in Arizona it was the Palo Verde project, in California it was Diablo Canyon, in Kansas it was Wolf Creek. Citizens in those areas, too, raised strong opposition to the enormous costs of the plants and the fact that their power was not needed. But because these plants were being built by private companies, they had little leverage for stopping construction or avoiding the cancellation payments, as citizens had in the Northwest.

As the total began to mount, *Forbes* magazine featured the failure of nuclear power on its cover, describing it as "the largest managerial disaster in business history," costing more than the nation's space program or the Vietnam War. All across the country, efforts to stop the crisis were piecemeal. There was no equitable way to resolve what had become a national dilemma. Efforts to phase in the rate increases over a longer period of time were expected to soften the blow of rateshock, but there was no prediction for what would happen as bills crept ever higher.

"So far," Komanoff said, "only the first two or three billion dollars has been split as a loss between consumers, investors, and taxpayers. This is only the tip of the iceberg. The question is who is going to pay the remaining $95 billion?" Events in the Northwest, he said, foreshadowed a larger struggle, pitting investors against consumers over who would ultimately absorb the debt.

The cause of this catastrophe was attributed to an honest mistake in projecting the growth needs of the industry. The traditional rate of expansion through the 1950s and 1960s was 7 percent per year. Basing its projections on this historical trend, the industry had undertaken its massive building program. As construction progressed and electric rates increased, consumers cut back on use, leaving the industry with partly completed plants and a growing power surplus. This is the traditional explanation, which is useful as far as it goes.

What it does not reveal is that the industry was doing everything possible in its advertising, marketing, and lobbying in the 1960s and 1970s to keep growth expanding at a maximum pace, including kickbacks to builders to construct all-electric homes and wholesale industrial rates that encouraged tremendous waste. As a result the industry built its future on vastly inflated electricity demand—a foundation of sand. Conservation and efficiency efforts advocated by consumer activists could have eliminated the need for the wave of construction. But regulators across the country ignored consumer proposals and sided with the power companies.

Ironically, the industry laid the blame for the soaring costs of the plants at the feet of regulatory agencies and consumer activists who intervened to stop the projects. The bottom line, however, rests on the fact that private power companies are allowed to charge rates based on the value of their assets. As they built more plants, rates and profits could go up, and as the cost of each plant rose, so did rates and profits. As a result most companies did little to stem the rising costs because to do so would mean cutting profits.

The nonprofit public and rural cooperative power systems, believing in the optimistic growth rates and fearing that they would be taken over by the expanding private power companies, launched their own building programs, such as the WPPSS project. Their mistake was that they did it without telling the public what they were doing.

On the surface, what is now at stake in this gigantic showdown over costs is whether or not the United States has reached a dead end with its nuclear program. But there is much more. The fight over nuclear power and the economic price it carries represents only the most visible part of a longer-term conflict, in which the stakes are higher and the less visible costs more disastrous than those we now recognize. The central issue is who decides what economic and environmental risks local communities will face in the trade-offs for increased electrification. In addition to soaring electric rates, the risks include annual billion dollar damage to crops and forests and lakes from acid rain, carbon dioxide build-up in the atmosphere (the "greenhouse effect") which is expected to lead to climatic changes, the effects of strip-mining millions of acres, transport of highly radioactive wastes on the nation's highways, and dumps to store the wastes for tens of thousands of years. The electric industry has a major impact on the nation's water resources: its thermoelectric plants use some 210 billions of gallons per day, almost half the total 450 billion gallons nationwide. The industry also carries the danger of nuclear weapons proliferation from nuclear reactor wastes, health risks from extra-high-voltage power lines, the growing vulnerability of centralized electric systems to sabotage by terrorists, and an erosion of democratic control of communities.

The political environment in which decisions about these and other risks have been made has long been heavily influenced by the industry.

When electricity first emerged from the back rooms of inventors like Charles Brush and Thomas Edison, it hit nineteenth-century America with a dazzling impact. What fire had been for early man was a rough draft for the force electricity took on in lighting cities, running hundreds of thousands of industrial motors, engendering extensive networks of trolley car lines, and sparking the birth of mass communications. Even more than the railroads of a few decades before, it quickly outstripped the understanding and control of social institutions. At the top of the industry were the equipment manufacturers, led by General Electric, which controlled the stock of hundreds of local power companies. Also tied in at the top levels were the major banks and investment houses that supplied the money for constant expansion, and the fuel companies that provided coal and later oil and uranium for the power use that doubled every decade. Together these companies made up the core of what was commonly recognized as an electric power empire bound together by a common belief that the prosperity of the nation, as well as their own success, was tied to maximum increase in the use of electricity.

The empire's influence ran from local city halls to the corridors of the federal government. Each power company was given a monopoly right to a geographic territory by a city or town. To protect those rights and expand control of power resources, such as the nation's rivers, the empire developed a system of deepening economic and political influence. Efforts by citizens to create public nonprofit power systems, which had been successful at the turn of the century, were steadily undermined by the private power monopolies and their corporate allies.

At the local level the private power companies were supported by industrialists, who received bargain rates subsidized by residential consumers. They were also supported by local banks, whose directors often sat on their boards and profited from the substantial borrowing done by the power companies. By the 1920s, the power empire consisted of a string of sixteen private holding companies that controlled nearly all of the nation's electricity. Through the National Electric Light Association these companies maintained a network of state and regional "committees for public information" that carried out extensive jointly funded propaganda and lobbying campaigns to elect political candidates, weaken attempts at regulation, and above all prevent public control of their monopolies.

Many political leaders were alarmed at the rise of the power industry's influence. Pennsylvania Governor Gifford Pinchot remarked in the early 1920s that "Nothing like this gigantic monopoly had ever appeared in the history of the world" in its ability to affect and control people's lives. He warned that if the industry went uncontrolled, it would be "a plague without previous example."

In the 1930s, the power empire tried to block the presidential nomination of Franklin D. Roosevelt, who had opposed their overcharges and political machinations while he was governor of New York. Roosevelt and the New Dealers succeeded in partially breaking up the power empire with strong regulation and support for city officials and groups of farmers to form their own nonprofit electric power systems. But the private companies would not be held back. In the late 1940s and early 1950s they pressed for commercialization of nuclear power as their vehicle to renew the power empire. Their drive to assure private control of atomic power in the early 1950s, played out in the atmosphere of the McCarthy era, was partly in response to a fear that the nonprofit municipal and rural cooperative power systems would gain access to nuclear power first and undermine their political and economic clout.

Under the Eisenhower Administration they regained their control over the industry, and by the early 1960s, the private companies began a new wave of centralization, creating new holding companies and promising an all-electric future. The plan was to triple the nation's electric production by 1980. Out of this plan for an expanded wave of electrification came the overbuilding and overfinancing and the crisis we now face.

Decade after decade for a full century, the guiding force of the industry was the struggle for control—the politics that shaped choices in both economics and technology.

In a strong and subtle way, the policies promoted by these corporate giants have pervasively affected life in the United States. Along with providing vitally needed electricity, they have bloated its consumption to keep sales expanding—resulting in the waste of enormous amounts of capital and unnecessary depletion of resources. Amory Lovins, a leading alternative energy advocate who has been labeled one of the world's "most influential energy thinkers" by *Newsweek* magazine, has likened some of the efforts of the industry to "friendly fascism." In testimony before a congressional subcommittee in June 1984, Lovins said that if energy efficiency planning had been implemented several decades ago, "our nation's successive energy crises would probably never have happened. . . ."

Today the electric industry is a patchwork of contiguous monopolies. It consists of 2,194 municipal and public utility district systems, which range in size from the city of Los Angeles to the hamlet of Readsboro, Vermont, 870 rural cooperatives strung throughout forty-six states, Puerto Rico, and the Virgin Islands, 210 private companies that supply power in most major cities, and six federal agencies that provide power to public and private systems from federal dams. These power systems are strung together by more than 600,000 miles of overhead transmission lines and some four million miles of distribution wires, carrying electricity from 10,499 power plants.

Unlike most major industrialized nations, where electricity is provided as a service by public agencies, the private companies are the dominating force in the United States, supplying 78 percent of the nation's electricity, controlling the transmission grids, and influencing much of the government's power policies. Within the transmission grids, the small municipal and rural electric systems are overshadowed by the private companies, and the federal power agencies are surrounded by them. State regulatory commissions, which were established in an attempt to control the private companies in the early 1900s, are highly influenced by their political clout.

Private power company executives reject the idea that their network makes up a new "power empire," but there is no denying the coordinated organization, lobbying, fund-raising, propaganda, and legal actions they undertake. At the center of this network is the Edison Electric Institute, the Washington-based trade association for the 210 private power companies, which conducts political analysis and lobbying and coordinates educational work. It was formed in the mid-1930s, following an extensive Federal Trade Commission investigation which discredited the industry's former hub, the National Electric Light Association. A spin-off of the institute is the Committee for Energy Awareness, an organization located a few blocks away, which conducts regional and national media campaigns promoting nuclear and coal power. Their well-polished ads are commonly seen on nightly television and in major magazines. Until recent controversy, all the activities of both organizations were supported largely by donations from power companies, which were charged to consumers.

Providing broad political backup and a link between the electric power systems and other corporate allies is the Atomic Industrial Forum based in Bethesda, Maryland. It is an association of nuclear manufacturers, financiers, construction contractors, and power companies. This group has been a highly influential force since its founding in the early 1950s and played a key role in clearing the way for rapid commercialization of nuclear technology.

A more recently formed arm of the industry is the North American Electric Reliability Council, a private organization that gathers the data provided by nine regional "councils" dominated by the private power companies. This group was formed in the wake of the blackouts of 1965 and 1967 to coordinate the planning and future growth of the industry. The nine councils maintain significant influence over the regional transmission networks and operation of power plants, setting baseline figures and industry consensus for the building of new plants and transmission lines. Additionally, the Electric Power Research Institute, an experimental engineering firm located in Palo Alto, is jointly operated by the industry as its technological research arm. Through these key organizations, the individual power systems pool their efforts to maintain monopoly control over the political and technological directions of electric supply in the United States.

The biggest secret is Wall Street's involvement in this empire. Producing and transmitting electricity is the most capital intensive industry in the world, and the power companies have long been known as the dividend machines of Wall Street. The private power companies traditionally issue half of all the new common industrial stock every year and absorb a third of all corporate financing. As much as 40 percent of a consumer's bill goes to pay for financing charges. Thus, a large number of major bankers and brokers such as Chase Manhattan and Merrill Lynch are affected by decisions concerning power companies. Half of the income of major investment bankers is estimated to come from financing private power companies.

In the process of encouraging and building for maximum growth, significant amounts of money are also channeled to major engineering and construction firms such as Bechtel Corporation, Combustion Engineering, and Stone and Webster, and to fuel companies such as Exxon and Peabody Coal, as well as to equipment manufacturers like GE and Westinghouse. The historical concept that still bonds these companies together is a belief that the nation's prosperity (as well as their own) is dependent on the growth of electricity production and use. The financial community, influential manufacturers, and builders play substantial roles in protecting the empire's interests.

In the Northwest, what citizens came up against when they tried to halt runaway construction and gain greater public control over power project financing was not an isolated case. Opposition from Wall Street firms, nuclear manufacturers, construction contractors, and other power companies is commonplace. In Maine, Massachusetts, California, and other states as well, the power industry and its corporate allies contributed tens of millions of dollars in the late 1970s and early 1980s to oppose state laws aimed at greater public control over the industry.

In addition to this legislative opposition, the power empire has been involved in much more questionable tactics. In the mid-1970s and early 1980s the industry intimidated critics in documented cases of surveillance, wiretaps, harassment, and infiltration of citizen groups in an attempt to dampen criticism. It has also wielded substantial influence over state and federal regulatory bodies—a problem that contributed substantially to the overbuilding of the 1970s and 1980s. Michael Johnson, a candid seventy-three-year-old Pennsylvania utility commissioner, has called the "capture" of utility commissions by private power companies "a problem of gigantic proportions." Similarly, at the federal level, Nuclear Regulatory Commissioner Victor Gilinsky told a reporter for the *Washington Post* that "enforcement of rules had been very lax" before the Three Mile Island accident. "Utilities were getting away with quite a lot," he said, "and others thought they could get away with more." He described the safety problems and financial collapse of the nuclear industry as "the chickens coming home to roost."

At the state and federal levels, the industry applies strong pressure to see that sympathetic individuals are appointed to regulatory commissions and that regulatory laws are kept to a minimum. Their requests are sweetened by campaign contributions to key government leaders. In the 1981–82 and 1983–84 election cycles, recorded contributions to members of Congress from nuclear manufacturers and power companies totaled $3.3 million. Joan Claybrook, a former Carter Administration official and president of Ralph Nader's Public Citizen organization, calls this system of donations "legalized bribes." A study done by Public Citizen showed House members who always voted for the industry received four times as much money as those who voted against it.

The fruits of this giving are evident in tax and regulatory policies passed by Congress. For the period of 1981–1983 tax loopholes allowed GE and nine major power companies each to average zero payments in corporate income taxes. Westinghouse and six other major power companies averaged less than 2 percent in their payments during the same period. Most of the power companies paid less than half of the average corporate taxes.

The contributions and lobbying affect technology as well. From 1980 to 1984 funding for solar programs was slashed by 87 percent, and funding for conservation programs was reduced to nearly nothing, while money for nuclear research and construction was increased. In the spring of 1985, the Treasury Department proposed eliminating tax subsidies for all energy companies. When the flurry of lobbying was over, the electric utility industry had maintained the bulk of its support (equivalent to $333 per household in 1984—$185 of that for nuclear power). The tax support for renewable energy and conservation (equivalent to $20 and $10 per household respectively) was scheduled to be eliminated. When questioned about how this would affect competition between solar and nuclear technologies, one Wall Street insider wryly observed that solar and other renewable energy technologies would be able to compete with nuclear when their manufacturers could provide an equal amount of campaign contributions.

The tax policies effectively determine which technologies will be commercially viable and which will be developed for use in the future. Undermining solar and conservation programs affects not only the long-term cost of energy, but also the environment, the competitiveness of American industries, and national security. A continued reliance on centralized technology, such as nuclear and coal, of course means continued centralization of the power industry and the need for substantial amount of capital from Wall Street, which will be funneled to construction contractors and fuel suppliers, thus keeping the money cycle in the empire going.

There is a small catch, however.

□ □ □

Witnessing the beginning of environmental degradation from acid rain and strip-mining, steadily climbing bills, and the failure of regulatory control, people in communities all across the country have staged staunch resistance to the power industry's expansion. During the decade of 1974 to 1984, literally hundreds of major confrontations took place in government hearing rooms, meeting halls, and on dusty construction sites. As the nation's dependence on electricity grows over the next decade and industry leaders continue to threaten brownouts, blackouts, and power rationing, a deepening clash may be unavoidable.

One Edison Electric Institute writer has observed, "Although another full-fledged Civil War is unthinkable, the ideological battle between the 'expansionists' and the 'small is beautiful' forces could tear the economy apart unless some effective consensus of attitudes can be developed." Many alternative energy advocates hope that the forbiddingly high cost of nuclear power and free market forces will create an opportunity for the development of cogeneration, energy efficiency, and small power production that will break up the monopoly control of the private utility companies. Given the long-standing political clout of the industry, that hope may go largely unfulfilled.

Edison Electric Institute, General Electric, Westinghouse, the Atomic Industrial Forum, and other major players are working to spark a "second coming" of nuclear power in the early 1990s. The Reagan Administration has thrown its support behind the plan out of a belief that the economic recovery will crash if a new wave of plants is not built. Energy Secretary John Herrington told an industry gathering in San Francisco in November 1985, "Nuclear power must be prepared to be the workhorse that, along with coal and others, will meet this challenge of our future electricity needs. As we move forward together to meet this challenge, I want all of you to know that we do so with a President of the United States and a Secretary of Energy that are irrevocably committed to nuclear energy as an option for our future."

Both Herrington and Nuclear Regulatory Commissioner Lando Zech, Jr., say that without the construction of two hundred coal and nuclear plants before the year 2000, the nation will see serious power shortages. Internal documents reveal the Department of Energy plans to limit citizens' ability to intervene, and to clear away other "institutional barriers" through the late 1980s. A new design of a "safe" light-water reactor is set for 1990. And a new breeder reactor is planned for demonstration by 1995 and commercialization by the year 2000. Even if the power shortages seen as driving this plan don't materialize, Herrington has another rationale. "America taught the world how to harness the atom," he says. "We developed the technology that the rest of the world has borrowed and we have no intention of presiding over the demise of the American nuclear industry."

Herrington's nuclear vision is one of two widely divergent blueprints for the future. It originated in a 1983 Department of Energy report titled *The Future of Electric Supply in America,* which advocates construction of as many as 438 new nuclear and coal plants by the year 2000. That report was published under Energy Secretary Herrington's predecessor, Donald Hodel, the same official who had helped engineer the financial disaster in the Northwest. Similar to the statements Hodel had made seven years earlier in the Northwest, the report threatens brownouts and blackouts if a new building program isn't initiated.

The other vision of the future was contained in *A Perspective on Utility Capacity,* a report put together by energy analysts at the Library of Congress in the fall and winter of 1983. Supporting a number of earlier studies, it states that no new plants are needed if effective programs of conservation and energy efficiency are undertaken.

The difference between the two reports is enough to light three quarters of the nation. It is also a difference of how more than a trillion dollars will be spent and whether a government rationale for continued monopoly control and a new wave of electrification will be put in place.

The industry's vision of the future includes the building of more all-electric homes and office buildings, the spread of climate-controlled malls and indoor shopping areas, proliferation of automated industry, electric cars and trains, and ultimately all-electric cities. The construction of plants advocated by the Department of Energy is seen as essential to feed this widespread electrification. The new wave of building would constitute the most expensive series of construction projects in history, requiring as much as $2 trillion in capital.

The industry's critics call it "a blueprint for disaster." Consumers would face soaring bills. Strip-mining would be accelerated, with some areas of the Southwest designated "national sacrifice areas." More networks of extra-high-voltage power lines would be built. A projected 10 percent increase in sulfur dioxide and 25 percent increase in nitrogen oxide from coal-burning plants would add to acid rain problems. And threats of sweeping climatic changes from the greenhouse effect would also be intensified, as would the already controversial problems of transporting and storing highly radioactive wastes and the potential for sabotage by terrorists.

The alternative vision coincident with the Library of Congress report is for a revolution in energy efficiency and an "energy transition" over the next twenty to fifty years to decentralized solar and renewable energy systems. The transition has three stages: plugging energy leaks with efficiency and conservation, utilizing existing technology to produce new power and extend the life of generating plants, and finally using decentralized solar systems for electricity generation.

Amory Lovins, one of the chief architects of the energy transition concept, says that in effect it is already happening. "Since 1979," he explains, "the U.S. got 100 times as much new energy from efficiency improvements as from expansion of supply." He cites the existence of a million solar homes, the use of cogeneration—industrial boilers used to make both electricity and steam for manufacturing processes—and the use of new conservation devices as being at the heart of the revolution. What had been considered an "iron link" between the nation's prosperity (measured by the Gross National Product) and electricity, he says, has been broken. And only the surface of new efficiency has been scratched.

Lovins estimates that the United States may have too many electric plants already. While the nation uses electricity for 13 percent of its energy supply, only 8 percent is required to meet the nation's actual needs. An average home in San Diego, for example, could install new electricity efficient bulbs and appliances and consume only a quarter of the electricity it currently uses. Variable drives on industrial electric motors could cut their electric consumption by one third; more efficient lighting could cut overall consumption for lights in half; and using building design, heat pumps, and alternative heating and cooling systems could reduce the need to use electricity for this purpose significantly. Through efficiency gains, Lovins argues, we could have electric expansion in necessary areas and still need no new plants. The nuclear expansion proposals he says are "a kind of last hurrah mentality raising its head." Looking back a decade from now he says, "We'll find that all this effort has gone to try and preserve a nuclear option which no longer exists."

Power industry officials accept only a part of Lovins' rationale, however. They say there will be increased efficiency, but not as much as he estimates, and that centralized power plant expansion is essential. Industry estimates for cogeneration, for example, are that it will meet 5 percent of the nation's electric supply. Alternative energy advocates estimate it could meet 20 percent.

If power companies must be pushed into development of these "least-cost" options, the shove would have to come from state regulatory agencies. But a study released in early 1986 found that only eight states had authorized the regulatory rules and laws to push alternative energy development. Beyond the slick advertising of the industry touting plans for energy efficiency and the bright hopes of alternative energy advocates, the "transition" could stall in the first stage of plugging the nation's energy leaks. In many areas a number of older power plants, whose usefulness could be extended, are being pulled off line and dismantled to bring down the level of power surplus and create a need for new plants. Further, the loss of solar and renewable-energy tax credits at the end of 1985 and cutbacks in federal funding for solar research are signs of how easily alternative and energy efficient technology development can be sidetracked.

Tina Hobson, executive director of the Solar Lobby in Washington, says that the continuation of nuclear subsidies and the elimination of subsidies for solar and conservation represents a "future betrayed." She told a gathering of congressional aides and press at a briefing on Capitol Hill in October 1985 that the nation would face increased oil imports and a steady erosion of the economy if options for energy efficiency and solar energy were pushed aside.

The struggle between these two diverging factions over blueprints for the future is the latest battle in the century-long turbulence surrounding electricity. The issue is not the free enterprise system, as some political flacks would make it out to be. The issue is over whether it makes sense for large monopolies to control a vital resource that has increasing impact on social, environmental, and economic policy. More than a debate on technology, it is a fundamental issue of democracy.

The essential question is, will the industry continue to consolidate its political and economic influence under newly created holding companies—and a new electric power empire—or will there be a growth of local public control over this vital resource? In economic terms it's a choice of whether electricity will be a commodity whose monopoly producers will use their influence to maximize sales and expansion, or whether electricity will be seen as a nonprofit service tied to least-cost policies, efficiency, and meeting of unstimulated, uninflated electric needs.

The choice will undoubtedly be shaped by public response to the nuclear disaster in the Soviet Union, predictions of dwindling oil production in the United States, and on-going events in the Middle East. But more important, the choice will depend on the outcome of the power struggle building between the public and industry. Already, action at the local level for greater public control and a redefinition of electricity is stirring.

Beyond the rebellion in the Northwest, one of the most vivid examples is on Long Island, where the Long Island Lighting Company, its corporate allies, and the federal government are pitted against local citizens and county and state officials. At the center of the dispute is the Shoreham nuclear plant, which sits like a $4.5 billion ship in a bottle—completed yet unable to operate because of questions over safety and cost. Evacuation from parts of the island in the event of an accident at the plant is viewed as impossible. In an emotional meeting in which the

Suffolk County legislature voted down its best available evacuation plans, Representative Wayne Prospect expressed the seriousness of the county's opposition, "We are traveling a road from which there is no retreat. If any official in this nation attempts to open Shoreham, Suffolk County should summon all of its resources, including police powers, to see that the Shoreham plant never sees the light of day."

If it comes on line, the plant threatens rate increases of 50 percent or more, and widespread fear of an accident. If it doesn't come on line, it will bankrupt the Long Island Lighting Company. In response to the quandary, a group of citizens and local business people have raised a long-brewing proposal for a public takeover of the company. Under such a scenario, electricity would be sold by a public nonprofit power company and the Shoreham plant would remain shut down. In January 1986 the idea was endorsed by Governor Mario Cuomo and the majority of the state senate.

Similar efforts for public control have surfaced in New Orleans and Chicago and many smaller cities and towns. In the case of Chicago, Mayor Harold Washington urged a public takeover as one way to save the city from the steady cost increases of Commonwealth Edison's Byron and Braidwood nuclear plants. He told a thirty-one-member Commission on Energy that "Chicago is at a juncture where it has to make some hard decisions in the energy field. . . . We won't be driven into paupery."

These are just a few of the attempts to gain local control and make decisions over technology with the opportunity for full consideration of economic and environmental risks. As the nation moves toward a high-tech economy, decisions about electricity production and use will become increasingly central economic and social issues.

What people like Dorothy Lindsey of Grays Harbor County, Washington, Wayne Prospect of Long Island, Mayor Harold Washington of Chicago, and Amory Lovins are up against is more than just the strategies of influential bankers, politicians, and industrial supporters of the power companies. They are up against glittering visions of a future long promoted by the power industry—all-electric homes, electric cars, automated industry, all-electric cities—visions that provide no sense of what it will take to produce this electricity. And perhaps most importantly they are up against the momentum of history that has carried those visions.

In many ways the struggles in local communities have come full circle. The bitter ideological debates, the financial double-dealing, the intrigue in political back rooms, and the efforts of people to stop schemes for massive geographic and political expansion of the power companies were in full bloom in the early 1900s. In those upheavals, going back more than a century, lie the roots of the crisis and the struggle for control of electricity the nation now faces.

2. *Behind the Miracle and the Myth*

I believe in municipal ownership of these monopolies because if you do not own them, they will in time own you. They will destroy your politics, corrupt your institutions and, finally, destroy your liberties.

Tom Johnson, Mayor of Cleveland
1901–1909

On the night of February 20, 1905, Tom Johnson stood at the high desk of the steamy city hall room and accused two city councilors, H. B. Dewar and F. W. Wilke, of receiving bribes from the Cleveland Electric Lighting Company (CELC). Claiming he had proof, Johnson asked for a council vote to have the two men indicted. Dewar shouted that it was a slanderous lie and demanded Johnson appear before a grand jury with his evidence. For the next several moments a stunned silence reigned as Johnson read out his charges. In addition to Dewar and Wilke, he also accused fifteen Republican members of the council of receiving campaign contributions from the electric company. As their names were read, some of those charged abruptly left the room. Behind the scenes a battle had been raging for months over the creation of a city-owned electric power system. Voters had approved a measure to create a city-owned power system in the fall election. But the CELC, which controlled most of the electricity in the city, had staged a bitter campaign of opposition and persuaded the council to vote against the measure. Johnson would later show how the company provided funds to Dewar and Wilke and the city Republican committee.

Johnson's allegations sprawled across the headlines of Cleveland newspapers, but such fiery conflicts between local government and electric companies were not uncommon at the turn of the century. It was a wild and woolly time, when business and government operated with few principles and public officials faced unlimited opportunities for personal gain. It was also an era of huge industrial mergers when today's giant corporations were first formed. In major cities across the nation, journalists and government investigators turned up a steady stream of city councilors for sale and government officials acting on behalf of private interests. One of the choicest plums of public office, and one rife with corruption, was the granting of franchises for electric power, the new industrial heartblood.

Tom Johnson knew the depth of corruption from his business experience in street railways and his service in government. The son of a destitute Confederate colonel, he had worked his way up from poverty by inventing an automatic coin fare box and later owning several streetcar lines. While serving in Congress, he was shocked by the corruption he witnessed, and he returned to Cleveland with a belief that reform of government had to begin at the local level. He claimed the people of Cleveland were being milked of millions of dollars by franchises given to streetcar and electric companies. He also pointed to the power companies' involvement in local politics. "They make a daily, hourly business of politics," he said, "raising up men in this ward or that, identifying them with their machines, promoting them from delegates to city convention to city offices. They are always at work protecting and building up a business interest that lives only through its political strength."

In his highly controversial election campaign in 1900, he promised a three-cent fare for streetcars and targeted the Cleveland Vapor Light Company and the Cleveland Electric Illuminating Company for public takeover. Through his decade-long fight to make good on these promises, Cleveland came to embody the early struggles over electric power that also emerged in San Francisco, Chicago, Detroit, and hundreds of smaller cities and towns. They were far-reaching conflicts, laying bare the extensive political and economic control the electric power industry was developing as it started on the road to becoming the nation's largest and most capital-intensive industry.

Contrary to popular myth, the development of electric systems didn't start in the United States. Serious experiments in electricity had been carried out in Europe as early as 1730. In England in 1746 electricity was being transmitted over iron wire two miles in length with voltage strong enough to kill small birds. In colonial America, Benjamin Franklin and others also experimented with electricity—referred to by the prominent Methodist preacher John Wesley as "the soul of the universe." But for decades the phenomenon remained a curio with no practical application.

During the industrial revolution the potential for electricity began to take shape. The electromagnet, the telegraph, and crude electric motors were developed by 1840. In 1853 La Société Générale d'Électricité was formed in Paris, and electric lighting made its way from the rooms of inventors and darkened exposition tents into lighthouses. In England the lighthouse at Blackwell was electrified in 1857, and the following year the South Foreland light had a generator installed. As the excitement over electricity spread through Europe, Jules Verne, the French science fiction writer, popularized its use in his book *Twenty Thousand Leagues Under the Sea,* published in 1860. In it, the *Nautilus* submarine was powered by electricity from a chemical process using ocean water or "sea coal." Onboard the ship were electric clocks, electric lights, and other devices. Out of the fervor in Europe the popular interest in electric lighting spread to the United States.

Part of what caught America's eye were public demonstrations held in both Paris and London. At the Grands Magasins du Louvre an installation of eighty arc lamps developed by the Russian inventor Paul Jablochkov dazzled spectators in 1877. A similar system was installed along the Thames Embankment in October 1878. Two months earlier, the Gaiety Theatre in London had been illuminated by a circuit of lights strung to an electric generator. In the United States, it was Cleveland, scene of Tom Johnson's fight for public ownership twenty years later, that was the first city to demonstrate an electric street lighting system.

In the post–Civil War decades, Cleveland had ballooned from a rural market center of 43,000 on the shore of Lake Erie to an industrial metropolis of 380,000. Its manufacturing firms pushed the city to first place in production of goods ranging from nuts and bolts to sewing machines. And in shipbuilding it surpassed even Philadelphia. In the midst of that explosive growth, Charles Brush, a local inventor, went before the Cleveland City Council in 1879 with a proposal to demonstrate his lighting system. With the generator or "dynamo" he invented in 1876 and arc lights he had perfected by 1878, Brush had been illuminating the offices of the Telegraph Supply Company on St. Clair Street where he worked on his experiments. He now wanted to light Cleveland's streets. The council was eager to witness the new technological wonder and set the lighting demonstration for Tuesday, April 29, 1879.

Newspapers in Cleveland and surrounding villages heralded the event far and wide. Late afternoon found crowds pressing into the city in mud-spattered wagons and the streets filled with a carnival atmosphere, bands playing and throngs of people on the sidewalks. On the public square a dozen ornamental poles fifteen feet high topped with large glass globes had been erected at intervals. Inside each globe was an arc lamp, two carbon rods that held a brilliant bow of current in the gap between them. The lamps were rated at 2,000 candle power, equivalent to floodlights. Amid rumors of their brilliance people carried clerks' eyeshades or smoked or colored glass to protect their sight from the "bright as day" lights.

As the sun set, thousands gathered around the square. Finally, in the spring darkness at 8:05, Brush gave the "go" signal. His assistant threw the switch to the dynamo a block away, and the first globe flickered with purplish light. Thundering cheers gave way to astonishment as the other lights came on. In the awed quiet the Cleveland Grays band struck up its brass, and artillery boomed along the shorefront. Cleveland became the first city to light its public square with electricity.

Brush's progress to this point had been rapid. He had sold his first light and dynamo in early 1878 to a doctor in Cincinnati, who dazzled his neighbors by using it to light his porch and yard. That October he had appeared with his arc lights at the Mechanics Fair in Boston, resulting in sales that marked several small milestones. On November 9, 1878, his lights flickered on at the Continental Clothing House on the corner of Howard and Washington streets in Boston. One light shining on the sidewalk outside the building was the first in the streets of the city and attracted crowds of curious people each evening. In nearby Providence, Brush sold eighty lights to a textile mill, the largest electric light system in the world. Later in September 1879, he sold two dynamos in San Francisco that gave a start to the California Electric Light Company, and the first central power station in the United States. By the end of 1879 his arc lighting systems brightened the long rooms of mills in Hartford and Lowell, and several dry goods stores in New York.

The demonstration on the Cleveland public square and his initial sales success solidified Brush's reputation and the prospects for broader marketing of electric lighting systems. With a little capital from the Telegraph Supply Company, the Cleveland inventor began operation of the Brush Electric Company in 1880. He developed a central plant near Ontario and St. Clair streets and began construction of a second on Lime Street. Within six months lights were wired on several of Cleveland's downtown streets. To replace the ornamental poles of the first demonstration, a towering steel mast was erected in the public square with eight arc lamps of 4,000 candle power at the top. Electric light service was provided to the city for the charge of a dollar per hour, doubling the light produced by traditional gas lamps, which were said to appear faded and jaundiced in comparison. While skeptics warned of the danger of fire and electrocution from the lighting systems, use of the technological miracle and the awe surrounding it continued to spread. By the end of the following year Brush central stations would be established in Boston, New York, and Philadelphia.

Brush was a front-runner, but he was by no means alone in his experimentation. In Connecticut, the firm of Wallace and Sons was also marketing a lighting system. In Newark, New Jersey, a young Englishman by the name of Edward Weston made a public display of his arc lighting system. In Philadelphia, Elihu Thomson and a partner, E. J. Houston, also developed a system by the spring of 1879 and sold it to a local bakery. And in dozens of other towns as well, local mechanics and inventors tried successfully and unsuccessfully to re-create generators and lighting equipment.

There was also thirty-two-year-old Thomas Edison, already world-renowned for his invention of the phonograph and his work on telegraph and telephone systems. The pragmatic Edison jumped into experiments with electric lighting in 1878 after the excitement over the demonstrations in Europe caught the attention of Grosvenor P. Lowrey, his patent attorney. Lowrey urged Edison to get into the business, and together they created the Edison Electric Light Company in late 1878.

While Brush was selling his arc lighting systems, Edison began experimenting and conducting demonstrations in the area around his Menlo Park, New Jersey, workshop. As Brush worked with little help designing, inventing, and marketing his lighting systems, Edison was well-financed and had a team of more than two dozen specialists at his laboratory. In contrast to Brush's brilliant arc lighting, what Edison wanted was "a candle that will give a pleasant light, not too intense, which can be turned off and on as easily as gas." The problem for his specialists was to find a filament that would grow hot enough to become incandescent, but would not burn out. The quest included highly publicized excursions into the jungles of South America. Innumerable exotic materials were used for a filament before, almost by accident (according to the myth), the team fell upon common cotton thread covered with lamp carbon. Actually, the British inventor Sir Joseph Swan had experimented with a similar carbon-covered filament inside a vacuum bulb as early as 1860. Edison, who was a voracious reader of other inventors' works, is commonly assumed to have borrowed from Swan's experiment. In the mythmaking that came to surround the industry, Edison's role was exaggerated to make his name synonymous with electricity. But perfecting the softly glowing filament in the incandescent bulb is his primary contribution to the technology of the power industry.

In late 1879 and early 1880, Edison and Lowrey staged widely publicized events at the Menlo Park workshop. Special trains from New York and elsewhere brought hundreds of people to view the Edison electric light. Following a demonstration on New Year's Eve 1879–80, a New York *Herald* reporter wrote that twenty lights burned in the street leading from the train depot to the laboratory. The lab itself was "brilliantly illuminated with 25 electric lamps."

Brush, too, was busy with his demonstrations. In March 1880 the city council of Wabash, Indiana, had Brush set up four 3,000-candle-power arc lamps on the courthouse dome, 200 feet from the ground. In a scene similar to what had occurred in Cleveland, 10,000 people converged on the Wabash town square on the night of March 31, 1880. When the lights came on in the moonless dark, suddenly illuminating the houses, yards, and distant river, an eyewitness reported that the crowd was "overwhelmed with awe . . . the strange weird light, exceeded in power only by the sun, rendered the square as light as midday. Men fell on their knees, groans were uttered at the sight and many were dumb with amazement."

Brush also illuminated a portion of Broadway from 14th to 26th streets with fifteen arc lamps on December 20, 1880. *The New York Times* reported, "the light flared out in the southwest corner of the square—4,000 feet from the generators in West 25th Street, and at the same instant every lamp along Broadway for three quarters of a mile appeared illuminated. The white dots broke out—not one after another—but instantaneously as though a long train of powder had been fired." Thousands of Christmas shoppers crowding the district were taken by surprise at the sight. Exclamations of admiration and approval were said to be heard on all sides. And a reporter wrote that people also speculated over the effect electric lighting would have on the gas companies and the possible downfall of these monopolies as a result.

As popular wonder over the miracle of electricity spread, competition with Edison became keen. Behind the miracle and the mythmaking, a harsh reality was brewing. On the same day Brush was staging his demonstration on Broadway, Edison wined and dined the Tammany Hall bosses of the New York City Board of Aldermen with champagne and food from Delmonico's at his Menlo Park lab. After dinner he treated them to a special demonstration of the lighting system. With the help of a twenty-one-year-old British administrative genius named Samuel Insull, who would later shape the destiny of the power industry, Edison was laying plans for development of a central power station and transmission system for New York City. Four months after the dinner, on April 19, 1881, he received a franchise from the aldermen to operate in the Wall

Street district, stomping ground of the railroad magnate and financier J. P. Morgan, who had invested heavily in Edison's idea.

While young Sam Insull traveled the country tracking down contracts and additional capital, The Edison Electric Illuminating Company (EEIC), the forerunner of Consolidated Edison of New York, began digging trenches to lay the wire safely from a plant site on Pearl Street to the first fifty buildings in the Wall Street district. On September 4, 1882, when the lights went on in the offices of Drexel, Morgan, and Company at 23 Wall Street, the press proclaimed the dawn of a new era. Soft, mellow light "grateful to the eye," the *New York Times* reported, would soon make daylight of the night. Even though it was preceded by electric systems in cities such as Cleveland, Wabash, and San Francisco in 1879 and 1880, the mythmaking behind Edison and popular rewriting of history made the Pearl Street plant the nation's first centralized power system.

Limited by their technology, those early power systems remained largely neighborhood affairs, which generated electricity on site for a single building or for just a few city blocks. Even the centralized plants, such as Brush's Lime Street station and Edison's Pearl Street operation, could generate power economically for a circuit of wire only a mile or so long. There was some public fear over the danger of electric distribution wires, but the low amperage 110-volt direct current that flowed on the lines was seen as relatively safe and would not kill anyone unless the hapless victim touched the wires while soaking wet. The generators or "dynamos" were huge. Out of Sam Insull's idolatry of P. T. Barnum and Edison's flair for capturing the popular imagination, the 6,500-pound dynamos at the Pearl Street plant were named after Barnum's circus elephants. Together the six generators produced enough electricity for 1,200 sixteen-candle lamps. They were connected to boilers rated at 240 horsepower, each of which devoured 5 tons of coal and 11,500 gallons of water a day.

With the highly visible success of the Pearl Street plant, Edison wanted to move on to larger centralized plants. While Brush was interested in selling individual power systems or providing the service at a fixed charge, Edison was the first to conceive of selling electricity itself. He saw the growth of the business as dependent on construction of central power stations, each generating and transmitting electric current to hundreds of thousands of customers, all paying operating companies year after year all their lives for a supply of electricity.

J. P. Morgan, however, disagreed. He preferred Edison to concentrate on manufacturing electrical equipment. He saw centralized power plants as costly investments that would be slow to return dividends, and as the largest shareholder in EEIC, he wanted to maximize profits. Morgan's concern at this early date illustrated what would be a conflict throughout the history of the industry, as investors pressed for technology to be selected on the basis of ever higher returns, often at the sacrifice of service to the public.

The growth of EEIC's individual lighting systems was rapid. By the spring of 1883, the company had 334 generators in operation, the sites ranging from the *Pittsburgh Times* building to cotton mills and grain elevators. The dazzling gem was Haverly's Theatre in Chicago. It contained 637 incandescent lights, including a chandelier and separate lights in the vestibules and dressing rooms, all run by power from two dynamos in the basement. Despite this success, Edison and Insull continued to fight with Morgan over the need for centralized plants. For young Sam Insull these disputes left a fateful hatred, setting the stage for disastrous conflicts with the Morgan family in the future.

Finally, defying Morgan, Edison put up his own money to push ahead with central stations. Within eighteen months, in a series of deals that must have surprised even Morgan, Insull made contracts to build central power stations in more than twenty cities and towns. At the time, Brush too had agents traveling from town to town, gathering local businessmen to form power companies. He sold equipment in Boston, Philadelphia, and Detroit, giving each company an exclusive right to a particular territory under the firm's patent. By 1884 street lighting in major American cities was commonplace. One British traveler wrote that it was a backward city that had not replaced some of its gas lamps with electric lights. The same phenomenon was also taking place abroad. With backing from British investors, both Brush and Edison purchased a number of foreign patents and introduced their lighting systems in Britain, Spain, France, Russia, and Italy.

The industry was quickly dividing into two parts. At the top were electrical equipment manufacturing companies dominated by Brush, Edison, a manufacturer of railroad air brakes by the name of George Westinghouse, and the Thomson-Houston Company of Lynn, Massachusetts. At the local level were the illuminating companies. As Morgan had predicted, the real money to be made at this early stage was in owning patents and manufacturing electric equipment. The local illuminating companies were commonly controlled by the manufacturing firms, which took one third of their stock in order to control the geographic market for equipment.

Despite its dazzling attraction the central station business was not as profitable as Edison had hoped. A severe economic depression was sweeping the country, and the demand for electricity in the early 1880s remained relatively low. A huge technical difficulty also lay in the fact that plants were utilized only part-time, primarily providing electricity for street lighting at night. The problem gave rise to a new avenue of development—the electrification of streetcar lines and railways.

In New Orleans, South Bend, and Indianapolis an inventor named Charles Van Depoele achieved a huge success establishing a series of electric streetcar lines in 1885. By 1890, fifty-one cities had service. Five years later electric trolley cars would be operating on 10,000 miles of track in 850 cities.

In the early 1880s, at the beginning of that burst of development, a number of electric illuminating company owners switched to the streetcar business, in which they could provide power for the cars during the day and power for street lights at night. Even the skeptics were swayed as they watched electric lights going on in the streets and streetcars powered by mysterious overhead wires.

Lighting for homes, however, was still limited. The cost of electricity in the early days was equivalent to twenty-five cents per kilowatt hour, roughly three or four times the average cost in 1986. For those wealthy enough to use electric lighting, the cost was about $100 a year, a quarter of the average annual income at that time. Many of the homes with electricity, such as J. P. Morgan's, had their own generators. The service from the central stations was marked by frequent blackouts, and a row of kerosene lamps was kept always trimmed and ready on the window.

Watching the early growth of combined streetcar and electric street-lighting systems, many entrepreneurs began to see the business potential Edison had dreamed of. Uncountable millions of dollars were to be made selling electricity. The keys to success were secure relationships with banks and Wall Street firms for a steady supply of capital, with manufacturers for credit and patent rights, with fuel companies, and with local politicians who granted the essential franchise to operate.

By the late 1880s tensions began to build in this infrastructure. As the number of central power stations multiplied and interest in the business grew, deep-seated conflicts began to emerge between competing companies and between the companies and city governments. In an age of horse-drawn streetcars, many regarded the franchises merely as concessions to perform a service. But in reality it meant a system of political favors in which corruption could be rank. With the coming of electric franchises, investment and competition increased—so did graft. This was especially true in major cities where a fat market was expected.

What was to happen with electric systems had been foreshadowed in scandals over the railroads. In the decade before the fledgling electric companies arrived, enormous sums of money were spent on politicians by railroad companies seeking land grants, tax relief, and subsidies. A disgruntled congressman from Ohio declared in 1873 that the "House of Representatives was like an auction room where more valuable considerations were disposed of under the speaker's hammer than any other place on earth." Investigations revealed that between 1866 and 1872 the Union Pacific Railroad alone spent $400,000 on bribes, and between 1875 and 1885 graft cost Central Pacific as much as $500,000 annually.

According to one observer in the late 1880s, nearly every city in the country had granted several general electric light franchises to competing electric companies. Out of a belief that maximum competition between the companies would keep charges low, some city governments granted franchises to all companies desiring to supply electric service. As a result these cities became ensnared in the wires of the new technology. Streets were often wired by one company and rewired by competing companies. In some commercial districts, where demand was high and competition was thick, forests of poles strung with wires appeared on the streets. In other districts where there was only demand for a few lights in each house, people had no access to electricity.

Vicious infighting between companies erupted. It resulted in increasing failure in service, higher costs, and gradual consolidation under companies that gained a political and economic upper hand. In Chicago, for example, a number of arc lighting ventures had mushroomed in various parts of the city and indulged in ruinous competition. None of these survived the 1880s. In 1887, a new company, the Chicago Arc Light and Power Company, was incorporated by the backers of the Gas Trust Company. They recognized the growing popularity of electricity and decided to join rather than fight the new enterprise. At once the company began to buy up the small rival arc lighting companies. Within a month the strategy was so successful that the Chicago *Tribune* observed, "Beginning this morning the sun and the Gas Trust has [*sic*] a monopoly of all the light with which Chicago is to be blessed."

In the late 1880s other cities were witnessing a consolidation similar to what was happening in Chicago. It was an ominous trend for those who had watched the centralization of railroads and their soaring rates and political scandals. Faced with the growing political and economic clout of these electric companies—whose service would come to saturate commercial businesses, industry, transportation, and homes—many local government leaders were frightened. Right from the beginning in some cities, far-seeing city fathers, such as those who witnessed Charles Brush's demonstration in Wabash, issued no franchise. Instead they established a city-owned electric system, providing power as a nonprofit service similar to water or gas service.

At the same time Edison was opening his Pearl Street plant, the city of Fairfield, Iowa, fired up a Brush dynamo and six 2,000-candle-power arc lamps as the beginning of a system that would provide electricity at nonprofit rates to its citizens. By 1888, fifty-three municipally owned power plants had been established in cities ranging from Lewiston, Maine, to Crete, Nebraska.

For the owners of private electric companies, the most disturbing fact about public power companies was that their charges for electricity were half that of the privately owned. In these towns common people gained access to the miracle of electric lights, while in other cities only the wealthy could afford to switch from traditional gas and kerosene lamps, or commercial businesses faced higher rates.

The brutal infighting between private power companies gave way to a wave of conflicts over whether a town government or private enterprise would control electricity. In major cities the debates raged behind meeting room doors and in public forums. It was the beginning of a far-reaching power struggle that would last for more than a century and come to have a deep impact on local communities, the nation's rivers and coal-rich lands, regional economies, and the nation's political atmosphere. On one side were those using Edison as their figurehead, insisting that free enterprise was the only way to develop electric power systems. On the other side were political reformers, aware of how far the control of the emerging electric trusts might eventually extend. At stake was the control not only of markets and geographic territories, but the expansion of political and economic influence, and ultimately the future of an industry to be worth hundreds of billions of dollars.

□ □ □

Reform mayors in San Francisco, Toledo, Chicago, and Detroit mounted intense campaigns to create public power systems in the late 1880s and early 1890s. In Detroit, Hazen Pingree, a prosperous shoe manufacturer and municipal reformer who would inspire Tom Johnson's efforts in Cleveland a decade later, was elected in 1889 after a campaign that publicized the corruption of Detroit officials by utility companies. Pingree promised to create a city-owned power system. "Good municipal government is an impracticality," Pingree told voters, "while valuable franchises are to be had and can be obtained by corrupt use of money. I believe the time has come for municipal ownership of street railway lines, water, gas, electric lighting, telephone, and other necessary conveniences which by their nature are monopolies."

What Pingree and the people of Detroit had witnessed was destructive competition between the local Brush Company and Detroit Electric Light and Power (DELP), a new rival company backed by Edison General Electric. When DELP began to string a duplicate distribution system, the local Brush Company staged nightly attacks to chop down poles. Although DELP brought this destruction to a halt with a court injunction, the battle continued. Each time a new contract was to be awarded, the rival firms spent money lavishly on city aldermen and provided heavy backing to their respective candidates in political campaigns. The dispute finally ended with a merger of the two companies in June 1891. As in Chicago, Detroit's power system fell into a single corporate hand.

Pingree railed at the swelling political influence of DELP and cited studies showing that power from municipally owned systems was being delivered in other cities at half the cost of electricity from private companies. He wanted a city-owned power system for Detroit, but General Electric balked at selling the necessary equipment to the city. Pingree then went to Western Electric and Westinghouse and received assurances that they were willing to sell generating and transmission equipment for a municipal plant.

In April 1893, Detroit citizens voted 15,282 to 1,745 on an advisory ballot in favor of a municipal plant. DELP and its parent company, GE, fought against the effort. William H. Fitzgerald, general manager of DELP, explained that the issue was larger than just Detroit. "If the city were to do its own lighting at about half what other companies bid, it would establish a bad precedent," he admitted, "and other cities that are now lighted by companies owned by the General Electric Company would be apt to follow Detroit's example."

DELP continued to woo city councilors who held authority over the creation of a city-owned power system, but with the weight of the voters behind him Pingree prevailed. He managed to establish a city-owned electric plant in 1895 to supply street lighting. True to his calculations, the cost of street lighting in Detroit declined from $132 a lamp per year in 1894 to $87 per lamp under municipal ownership in 1898 and to $63 by 1902. A short time later he urged the city council to extend service to stores and homes of private citizens. "If this were done," he declared, "it will take electric lights out of the luxuries of life, only to be used by the wealthy, and place it within the reach of the humblest of citizens."

The resistance that Pingree ran up against to provide the common people with electric power as a nonprofit sevice also existed elsewhere. In Massachusetts, for example, a home-rule petition from the town of Danvers, asking the state legislature for the right to sell electricity to local businesses and homes, led to a three-year legislative battle over the rights of municipally owned electrical systems. The legislature also received petitions from Boston, Worcester, Lowell, Lawrence, and a number of smaller communities, requesting the right to produce and distribute gas or electricity to their residents.

Patrick Collins, an attorney for the Boston Gas Light Company, called the proposal an "excursion into the dark socialistic jungle." Opposed by a group of sixty-one private gas and electric companies, what finally emerged in 1891 was a highly restrictive version of the bill. It had a chilling effect on public power in Massachusetts and in other states where it was copied. A city or town could not develop a competing system, but had to buy out any existing private utility. To do this a two-thirds vote of a city council was required in two consecutive years, followed by a majority approval of the town's residents. In the intervening period private companies could mount intensive publicity campaigns. The company could also stall for several years after a conclusive vote, while haggling over the fair market value of the system. During the next decade only seventeen additional municipal systems were established in Massachusetts. Of these, seven were located in communities which had no previous electric service.

The issue over public power versus private power continued to rage through the 1890s, moving from the local level to the floor of state legislatures. Cities all across the nation were now witnessing fights over who would control this vital resource.

Despite claims that the notion of publicly owned electric systems was "socialistic," it was not a new idea. The historic roots of public ownership can be traced back to a period before the American Revolution when the first municipal water systems were established. The central notion was that certain essential resources were best utilized according to common priorities. Access to clean, reliable water supplies was fundamental to the physical health of a community. Public-minded individuals in the nineteenth century saw electricity as the same kind of resource. A flow of cheap, reliable electricity offered a boon to the economic health of a town's businesses and industries, as well as to its citizens.

In Great Britain, where electrification of cities and towns was also evolving, similar questions had arisen. In 1879 the House of Commons was faced with thirty-four private bills regarding streets to be broken up so that wires for electric systems could be laid. Parliament subsequently passed a law providing private companies with the right to lay the wires, subject to approval of municipal governments. Town governments were also given preference rights to develop their own power systems first, as well as revisionary rights over the private systems.

In the United States the issue hadn't yet reached Congress. Focused at the state and local levels, the movement for municipal ownership of power systems became part of an effort to eliminate rampant corruption in municipal government. During the fight over the Massachusetts public power proposal, Richard T. Ely, an economist from Johns Hopkins University, connected political corruption with the rise of private corporations in control of monopoly businesses like electricity. In an address before the Boston Merchants Association in January 1889, he told businessmen that in addition to lowering the cost for power, municipal management of power systems would bring about "a harmony of public and private interests, awaken public spirit, and create a reformation in political life." Other academicians, such as John R. Commons of the University of Wisconsin and Edward Bemis and Frank Parsons of Boston University, provided reams of statistics to show that publicly owned power systems were more efficient and cheaper.

The ideas of many social reformers were sparked by Edward Bellamy's utopian novel *Looking Backward.* The novel's main character, Julian West, is hypnotized into a deep sleep in 1887 and awakens in the year 2000. Risen on the foundations of the industrial smoke-stained city of his day is an ideal community. Bellamy believed the Golden Age he depicted would arrive after the giant corporate trusts were nationalized. Along the way municipalities would assume control over local public services such as transportation, lighting, heating, and water supply. Bellamy called his system "nationalism."

Some 125,000 copies were sold in the first two years after publication, and "nationalist" clubs rapidly formed around Bellamy's vision. By 1890 there were 150 of these clubs in twenty-seven states. The Boston club claimed to be instrumental in getting the limited public power law passed in Massachusetts in 1891.

As reformers pushed for public power systems, there were other forces at work shaping broad consolidations in the electric industry. In 1888 the railroad magnate Henry Villard attempted to form a worldwide electric industry cartel backed by German investors. He offered Edison, who was continually at odds with J. P. Morgan, $500,000 to take part in the new conglomerate. Morgan made a counterproposal for a compromise among the three men, and out of their deal Edison General Electric was formed in January 1889. The board consisted of four Edison men and five appointed by Villard and Morgan. It pulled together Edison's six enterprises and held his patents.

The compromise lasted only a few years, however. In April 1892 Morgan, the lead stockholder in Edison General Electric, pulled the reins tighter on the industry by consolidating with the Thomson-Houston electrical company, which had swallowed Brush a few years before. Morgan then bought out Edison, who left the electricity business forever. Ultimately he also forced Villard to resign. The new company emerging under Morgan's control was to be known as General Electric.

That same year, because of a severe stock market crunch, consolidation of more than a hundred electric illuminating companies occurred. Because the manufacturing companies had been taking one third of the common stock of electric systems as part of their payment, the merger and the crunch meant that the new company, GE, controlled more than 1,245 central power stations and 2,300 individual lighting generators. Amid social upheaval and labor strikes, this growing concentration of the industry was the counterweight to the growing desire for publicly owned systems. It marked the beginning of a new and fateful era for the power industry.

Another development also fit neatly into the growing concentration of the power industry. In the early 1890s the central power stations were no longer limited to generating power only for their immediate neighborhoods. In 1886, George Westinghouse, the founder of Westinghouse manufacturing, had begun marketing a high voltage alternating current system that allowed transmission over longer distances. By 1892, Southern California Edison opened a transmission line for delivery of 10,000 volts to a site twenty-eight miles from the generating plant. In 1896, an electric power plant at Niagara Falls began supplying power to the Buffalo Street Railway twenty-two miles away. And by 1902 electric power was being transmitted as far as 200 miles in the San Francisco area. While it was more dangerous than the old low voltage direct current Brush and Edison began with, the high voltage alternating power allowed the size of the companies' territory to expand.

The development of large steam turbines to power the increased loads followed. In 1901, the Hartford Electric Light Company became the first American utility to install a steam turbine, which produced 2 megawatts of electricity, 2,200 times as powerful as Edison's Pearl Street Station. Two years later Commonwealth Edison of Chicago more than doubled this by installing a 5-megawatt generator. And by the end of the decade a 35-megawatt station was in operation in Chicago.

The economics of scale for central stations and long transmission lines ensured growing centralization of the power companies. This accelerated consolidation under the private companies further. For the social reformers and public power advocates, the steady growth in size of the private companies spelled trouble. Town-owned systems, limited by law to operate within city boundaries and without the larger consumer market or access to capital to build huge generating stations, would face growing pressure from private power owners.

At the turn of the century, the power industry continued to evolve very quickly. While Pingree had been relatively successful in creating a city-owned electric system in Detroit, what Tom Johnson came up against in his fiery city council battles in Cleveland ten years later was a company that had enlarged far beyond its local proportions.

In the two decades that had passed since Charles Brush's dazzling demonstration on the Cleveland public square, the infrastructure and political power of the industry was substantially expanded. Consolidation of small private illuminating companies had given rise to a larger and more powerful form of electric company—one connected to a regional or national holding company, whose board of directors often interlocked with those of banks, brokerage firms, insurance companies, or other interests with whom a city's officials had to do business. The CELC, which had been started by Brush, was now one of GE's subsidiaries. Behind it stood J. P. Morgan, the most powerful and shrewd businessman in the nation. When Tom Johnson proposed a takeover of CELC, he was talking about taking over a subsidiary of GE and fighting Morgan; no easy task.

After being rejected by the city council in February 1905, Johnson brought the issue before the voters again in the form of a proposal to annex the suburb of South Brooklyn and its municipal plant. CELC and its allies lobbied hard, but Johnson could not be stopped. The public vote carried a second time, and the city council was forced to accept the municipal plant in the suburb of South Brooklyn as the first step toward a municipal power system for Cleveland. The fight had taken ten years.

In 1911, after Johnson had left office, the citizens approved a $2 million bond issue to build a huge electric plant with four large turbines and a 24-megawatt capacity on the shore of Lake Erie. Failing to stop the new plant with court injunctions, Morgan tried to poison the city's effort to raise money for construction. In New York, Morgan's financial firms issued circulars advising investors not to buy Cleveland's bonds for the plant, but he was unsuccessful.

When the new Cleveland municipal system began operation in 1914 and established a rate of three cents per kilowatt hour, CELC was charging ten cents. It was the first example of a major city selling its own power and providing reliable service at a drastically reduced rate. As such a model, the Cleveland municipal system continued to draw attacks from GE and its financial affiliates.

Elsewhere across the nation, efforts at municipal reform and desire for cheap power fostered a remarkable growth of public power systems. Each year during the period of 1897 to 1907 between 60 and 120 new public systems were formed by referenda. The rate of increase for public systems was more than twice as fast as private companies. By 1912 there were 1,737 public power systems and 3,659 private companies in operation.

As the surge of growth in public power systems mounted, it frightened leaders of the private power companies, who would not sit idly by as their grip on the future of the industry and the political and economic control attached to it gradually slipped away. Bold new proposals were about to emerge to stem the trend for public control of electricity.

□ □ □

The new approach was state regulation. In 1898 Samuel Insull, Edison's former administrative genius, first proposed the idea at a historic power industry convention in Alabama. Insull was then forty years old —the aging boy wonder of the industry. At twenty-one he had arrived from England to serve as Edison's administrative secretary. By his mid-twenties he had become a key manager for Edison's companies, steering them with a corporate recklessness and an unerring memory for detail that made him an impeccable master at organization and sales. When J. P. Morgan had formed GE in 1892 and pushed Edison out of the business, Insull turned his back on Morgan's offers and took a job as president of Chicago Edison. He was thirty-four years old. Six years later, he owned the company and was on his way to controlling a fateful network of power systems that would span most of the nation. He stood before the National Electric Light Association (NELA)—an organization formed in 1885 to unify the industry's power policies—as its president.

Much to their amazement, Insull told the gathered executives that competition was "economically wrong" for the electric business. What power company officials had been reluctant to admit was that erecting duplicate lines and constructing duplicate power plants drove costs sky-high. Taking the shortcuts necessary to bring these costs down and compete made service unreliable. Insull pointed out that the business of delivering electricity was in fact a "natural monopoly." "While it is not supposed to be popular to speak of exclusive franchises," he told the group, "it should be recognized that the best service at the lowest possible price can only be obtained . . . by exclusive control of a given territory being placed in the hands of one undertaking."

His listeners were stunned by Insull's open admission that competition didn't work. For many it sounded like the death knell for private ownership of the industry. But Insull had a plan. They sat in disbelief as he outlined a proposal for state governmental agencies to fix rates and standards of service and insure exclusive control of a territory for a single company. Along with stabilizing tumultuous competition between companies, Insull believed the plan for state regulation would head off the growth of publicly owned systems by offering an alternative form of public control over the electric business.

Utility executives were appalled by the idea. In accepting a monopoly franchise, it would be necessary to give the state regulatory agency the right to limit their profits. Trusting in his experience, Insull was confident of the political magic built into his proposal. He had already bought up the competitors of Chicago Edison and at the time was in near total control of electric power in the city. Early regulatory proceedings before quasi-judicial commissions over franchise questions with city government had shown that the private companies possessed vastly superior technical and financial resources to win a case and maintain control of their rates and profits. As other power company executives became convinced of this, the proposal picked up steam.

In 1905 the National Civic Federation—one of the most important organizations of its day—whose roster included Insull, Andrew Carnegie, Consolidated Gas President George B. Cortelyou, and several partners in J. P. Morgan and Company, established a Commission on Public Ownership. The commission spent two years studying the power industry and published its findings in a three volume report in 1907. Given the presence of such men as Boston Edison President Charles Edgar and Walter J. Clark of GE on the study group, it was no surprise that the findings coincided exactly with the views of Samuel Insull and the principles of state regulation worked out by the NELA.

The commission called for a legalized system of electric monopolies with public regulation and examination under a system of uniform records and accounting. No position was taken on the general expediency of either private or public ownership. The report stated that the question of ownership must be resolved by each community in light of local conditions. Delegates at the 1907 NELA annual meeting were urged to lobby for the proposal and were told that implementation of regulation would reduce "the necessity or excuse for municipal ownership by securing fair treatment for the public."

Using the report's general framework for regulatory laws, Wisconsin and New York led the way in 1907 when they established utility commissions with general power of supervision over electric companies, which previously had been subject to local control. Before the close of the decade, similar laws were enacted in Vermont, New Jersey, and Maryland. And by 1921, all but Delaware had established state commissions with power to regulate utilities. The main purpose of the commonly understaffed and underfunded regulatory agencies was to substitute for competition and carry out the contradictory mission of protecting the interests of both the companies and consumers.

Political reformers were split over the prospects of controlling power companies through state regulation, which played into even more turbulent politics than those evident at the local level. Milwaukee Mayor Daniel Hoan said in 1907, "No shrewder piece of political humbuggery and downright fraud has ever been placed upon the statute books. It's supposed to be legislation for the people. In fact, it's legislation for the power oligarchy."

For home-rule advocates who favored control through municipal franchises, regulation by an appointed state commission was a disastrous mistake. Stiles P. Jones, a utility expert associated with the National Municipal League, an organization advocating betterment of local and state government, argued that the heated political struggles over local regulation of utilities were signs of a healthy democratic process. Keeping public officials on the firing line included "a proper appreciation of their duty. Local regulation educated the public in municipal democracy," Stiles said, and taught citizens self-reliance and a capacity for self-government.

The inequities in state regulation soon became apparent. As Insull had predicted, it was impossible for consumers to compete with private companies. The sky was the limit as to how much a company could spend on a case, and all expenses would eventually be charged to consumers in their rates. In the early days the cost of carrying through with a complaint normally ran anywhere from $10,000 to hundreds of thousands of dollars. In these proceedings city attorneys proved to be no match for specialized utility counsel brought in from other parts of the country. The utilities were also able to secure and pay technical experts who were reluctant to testify for the public side because, as one consulting economist candidly explained, "engineers and accountants experienced in electric lighting . . . are in the employ of private companies or expect to be."

In addition to a stacked deck in the hearing process, the state commissioners, appointed by a governor in most cases, were also subject to political pressure and often remained reluctant to exercise authority except in the most extreme cases. In the great majority of cases, rates went up or down through friendly, informal negotiations between commissions and company management. When an electric company disagreed with a commission's decision, the decision could be appealed to the courts, where the case could languish for years, thus crippling any effective public authority or control. Between 1907 and 1912, of ninety-one cases heard by the Wisconsin commission, the power companies were granted increases in fifty of the fifty-two cases they filed. Wisconsin cities and towns asked for reduction of rates in thirty-nine cases, which were granted in only eleven.

Nevertheless, in the first two decades of this century, as the states established utility commissions, there was great hope for the "scientific method" of regulation and an era of public control which would be "definite, precise, and eventually automatic." By the mid-1920s it would become apparent that the process wasn't working. Through their increasing political influence, the private utility companies had been able to achieve a major political victory. The establishment of state regulation protected their monopolies, institutionalized their economic and political influence at the state level, and allowed the beginning of massive growth of their companies and systematic plundering of consumers, and later of unwitting investors who would lose heavily in watered stock dealings.

Most importantly, state regulation helped slow the growth of public power systems between 1907 and 1917 to half what it had been during the preceding ten years. With their rationale for a "natural monopoly" secure, the private power systems used their growing size and economic weight to move to a second level of consolidation and centralization of the industry.

After the advent of the natural-monopoly concept and the institution of state regulation, the new technology of large-scale centralized generation and long distance transmission became increasingly common. By 1910, as power companies began to transmit electricity hundreds of miles, power lines began to appear on the landscape. The political change was spurring a profound transition in the structure of the industry. Small investor-owned utilities that could not compete with the economics of large companies were absorbed into increasingly integrated and interconnected systems. Insull told an audience of engineers in Dayton, Ohio, in 1914 that streetcar companies could make greater profits by shutting down their individual power plants and buying power from Chicago Electric. At that time he was in the midst of setting out the cornerstone of an ill-fated fiefdom.

Making Chicago his base, Insull adapted the holding company structure pioneered by GE and other manufacturers to control a number of operating companies and their markets. He began to expand outside the Chicago area in 1902–3 by acquiring electric light companies in four small towns that he strung into the North Shore Electric Company. As evidence of his growing influence, Insull appeared privately before the joint committee of the Illinois Legislature to push for state regulation. A bill was passed in 1913. Fully anticipating that event in 1912, Insull took his first major step by creating a new parent company, Middle West Utilities. In a financial sleight of hand, Insull inflated the value of his operating companies by a factor of 10 when he transferred them to his new parent company. In consideration of that inflated value, he could release more stocks and bonds and charge higher prices to consumers. It was a perfect example of the kind of stock watering that would later shatter the industry.

Insull was to use this process repeatedly. By 1916 he controlled 118 power systems operating in nine states, as well as steadily rising layers of power companies, subregional and regional holding companies with Middle West Utilities at the top. Following GE's and Insull's lead, similar holding companies began forming out of the power industry's financial and manufacturing infrastructure. Electric Bond and Share, a service company subsidiary of General Electric that had been established in 1905, coordinated extensive growth of GE's holding companies under the leadership of Sidney Z. Mitchell. Mitchell organized three subholding companies: American Gas and Electric Company operating in Ohio, Illinois, and Kentucky in 1906; American Power and Light based in Kansas in 1909; and Pacific Power and Light combining power companies in Washington, Oregon, and Idaho in 1910. In the background was J. P. Morgan as main stockholder in GE and the chief banker and broker for these companies on Wall Street. About the same time a young attorney by the name of Howard C. Hopson began learning, in college and in work for the New York regulatory commission, how private utilities were capitalized. Later, he would take that knowledge and enter the electric utility business with the purchase of Associated Gas and Electric

Company in 1921. By 1925, Hopson would control some 250 operating firms which provided steam, water, ice, and transportation to some 20 million people in 26 states, Canada, and the Philippines. Other companies, such as the Stone and Webster construction and engineering firm, also began to operate a growing number of power companies. But the main battle looming in the future was to be between Morgan and Insull as their expanding territories and ambitions began to collide.

Regulatory agencies and public power systems faced increasing difficulties as the holding companies expanded. Insull, Morgan, and others found that the highest profits to be made were not in the operation of power systems, but in providing financing, engineering, fuel, and equipment. The greatest cost of electricity was embedded in the company's financing charges. S. Z. Mitchell called this the "meat and bread" of the industry's owners. By bloating the value of the power companies in questionable stock deals, bankers and stock firms who owned the holding companies made huge profits that could be passed on as higher prices to consumers. The favor and support of industrialists was gained by granting them cheap rates. This added to the power companies' clout. For the new state regulatory agencies, it became nearly impossible to gain access to accounting figures, which faded from one set of books into another as power company operations were carried across state lines.

Both small independent private power companies and public power systems, which were limited to operate within city boundaries, faced the pressure of buy-outs. In order to force some public systems to sell out, private companies would start a rate war. Soon after the city of Pasadena, California, began generating its own commercial power in July 1907, it found itself engaged in a competitive rate war with Southern California Edison, which had refused the city's offer to buy its local system. The city entered the commercial electric business with a charge of eight cents per kilowatt hour. At that time Edison was charging fifteen cents. The tug of war for customers continued over the next several years until the city was forced to charge five cents to Edison's four cents. Because of the loud chorus of complaints and the fact that Edison was supporting its reductions by charging higher rates to surrounding communities, the California legislature took action. In 1913 the state passed the Unjust Competition Act prohibiting a company from subsidizing discounted rates for electricity in one community with higher rates in another community. The law made it impossible for Edison to force the Pasadena system out of existence.

At that same time, other state legislatures turned down proposals to allow expansion of municipal service outside of municipal boundaries as unfair competition with private companies. The municipally owned utility in Watertown, New York, for example, tried for two years to obtain permission to sell power to surrounding communities but the proposal was blocked in the state legislature. After a long fight, the city of Hagerstown, Maryland, was the first to receive approval from its state legislature to expand, but the private utilities had the Maryland Court of Appeals set the law aside on the grounds that it applied only to Hagerstown and was therefore discriminatory.

Failing to take over a public system, private companies often sought to capture it by offering wholesale power at a price that undermined the reason to build or operate city generating plants. This was a trend that would sweep the country, as private companies used their regulatory protection and heavy financial backing to extend their distribution and increase the size of their generating plants. In Anaheim, California, the municipally owned plant generated power until 1916 when equipment deterioration, increasing demand, and rising fuel prices made production costs too prohibitive. Rather than buy new equipment, the city signed a contract with Southern California Edison to purchase power.

The trend for this kind of switch was fairly rapid. In 1909, less than 7 percent of the 1,414 publicly owned systems in the United States purchased their total output from private companies. By 1923 the figure had climbed to more than one third. And in the state of Nebraska, 50 percent of the municipal systems were purchasing all of their power from private companies.

In addition to expanding their territory, the growing private power conglomerates also sought to gain control of as many potential power sources as they could. The advent of long distance transmission lines allowed them to move into the back country and develop dams or lay claims to future dam sites. While electric power was primarily generated from coal, approximately one third of the electricity generated in 1912 was from water power or "white coal." In a dispute that began with Teddy Roosevelt and raged through four presidential administrations, the early conservationists led by Gifford Pinchot, Roosevelt's chief forester who would later be a leader in the fight for public power, tried to prevent private power companies from seizing control of the nation's rivers.

Already in 1912, private corporations led by Stone and Webster controlled two thirds of the nation's water power. At the time, Congress maintained authority over the nation's rivers. Under the expanding political and economic influence of the holding companies, congressmen were conducting a giveaway of rights to the rivers. Roosevelt attacked the process and claimed that the "great corporations were acting with foresight, singleness of purpose and vigor to control the water power of the country." This would not only affect electricity production, but river shipping and irrigation for many of the nation's farms, especially Western and Midwestern regions, where water was critical to the local economy. Roosevelt believed it was his duty to "use every endeavor to prevent this growing monopoly, the most threatening that has ever appeared, from being fastened upon the people of this nation."

Long after Teddy Roosevelt left office and after nearly two decades of argument in Congress, a compromise would finally be reached with formation of the Federal Power Commission in 1920. The controversy over who would control the rivers was resolved by granting thirty- to fifty-year dam licenses to companies and allowing public power systems preference to sites where there was competition with a private company. The underlying conflict over public versus private power had churned to the surface of national politics. It was no longer a series of local or statewide conflicts decided in gilded back rooms. The confrontations that had erupted in the city council chambers of Cleveland and Detroit now took on national proportions. In the years before World War I, the struggle over water rights forced the federal government to take up the issue of public or private electric power.

Faced with the uproar at the local level and private monopolization of natural resources, Gifford Pinchot and the early conservationists joined with social reformers and local government leaders in calling for public ownership of the electric industry. In 1913 Pinchot and George Norris, the progressive Republican Senator from Nebraska, formed the National Popular Government League (NPGL). This progressive, nonpartisan organization lobbied for legislative measures to increase the powers of self-government. The bushy-browed Norris, who favored complete government ownership of the power industry, would prove to be private power's most tenacious foe in Congress. In defending the rights of democracy and standing up for consumers, he was continuously attacked by private corporations as a "socialist" and "communist." In the decades ahead Norris would be instrumental in restructuring the industry and urging the federal government to develop power agencies such as the Tennessee Valley Authority and the Rural Electrification Administration. With the league, he initiated a campaign for state legislation to allow greater use of initiatives and referenda designed to place more political power in the hands of the people. This effort toward greater self-determination was hoped to be a new source of strength for public power.

Judson King, a former aide to Toledo reform mayor "Golden Rule"

Jones, was appointed as the league's first director. During the hot summer of 1916 King and his wife drove all across the nation in a bone-jolting trip to see if the league's proposals were having an impact at the local level. They found that twenty-one states had adopted constitutional amendments allowing initiatives, but the majority were only marginally useful.

While the NPGL would continue to work on issues before Congress, it was obvious, if public power was to be defended and promoted, more work would be needed at the local level. That same year Pinchot helped organize another group, the Public Ownership League, (POL), aimed at maintaining the growth of local public power. At the opening meeting of the POL in Chicago were Pinchot, the crusading social reformer Jane Addams, Senator Robert La Follette of Wisconsin, Frank Walsh, who later served as the first chairman of the Power Authority of the State of New York, and a young professor by the name of Scott Nearing, who with his wife Helen would become venerated decades later for their back-to-the-land lifestyle and philosophy of self-sufficiency. Carl D. Thompson, a Wisconsin minister and member of the state legislature, was named secretary of the organization, a post he would hold for the remaining thirty-three years of his life.

The purpose of the organization was to help counter the mounting attacks on public power systems and protect local government from the growth of monopolistic forces that would run rampant over society's needs. The membership, which grew from 100 in 1916 to over 5,000 by 1922, believed in the public ownership of natural resources and "those public utilities which are natural and essential monopolies—water, light, power and transportation." The organization stood guard almost literally night and day over the next several decades to protect and promote municipal utilities and publicly developed power resources. Carl D. Thompson, perhaps the least known and clearly the driving force behind the organization, devoted his life to traveling the dusty back roads of the country lecturing and campaigning on the virtues of public power.

With a meager budget that began at $2,500 in 1917 and rarely rose over $12,000 in any one year, Thompson and his colleagues took on the growing power monopolies.

□ □ □

The 1920s were the golden years for electric power. The holding companies moved to a third stage of consolidation and mergers. By the middle of the decade, Insull and fifteen other holding company leaders controlled 85 percent of the nation's electricity. It was a new industrial empire—a power empire. Protected by state regulation, rising political influence, and access to nearly unlimited amounts of capital, they had smaller private companies and public power systems in a box. In the surge of mergers between 1922 and 1927, more than three hundred small private companies were swallowed each year by the holding companies. The growth of public power systems was brought to a standstill in 1923 when 3,066 public systems served one out of every eight electricity consumers. The drop in the next four years was precipitous, with a loss of 746 public systems.

Public power advocates like Gifford Pinchot, George Norris, and Carl D. Thompson watched as private utility executives used "every known method except sky-writing" in rate wars, buy-out campaigns, harassing litigation, and a massive propaganda campaign to undermine public power systems. Pinchot, Teddy Roosevelt's former chief forester and governor of Pennsylvania in the early 1920s, observed, "Nothing like this gigantic monopoly has ever appeared in the history of the world. Nothing has been imagined before that remotely approaches it in the thoroughgoing, intimate, unceasing control it may exercise over the daily life of every human being within the web of its wires."

Pinchot and Thompson recognized that the public power systems were hopelessly handicapped by their confinement within city boundaries and their difficulty in raising money for new plant equipment. In response they launched efforts to bring the small public power systems into a large integrated transmission system. Pinchot tried to develop a "Giant Power" transmission system within Pennsylvania that was bitterly fought by the private companies. Thompson promoted a nationwide "Superpower" transmission system linked to federal hydroelectric sites. With such networks they believed public power systems would be able to survive and prosper and act as a measure to keep the holding companies in check.

In 1924, after the holding companies had turned down the idea of a Superpower system to help bring down costs for consumers, Carl D. Thompson brought the concept to a National Public Ownership Conference as an option for public systems. Thompson urged the formation of all available electric power in the United States into one publicly owned system. He proposed the 2,318 municipally owned electric utilities as the starting points to be linked into a new national transmission system.

Under Thompson's plan, the federal government would build new generating plants and lay out main trunk lines, and local systems would fill in the gaps so electricity could be distributed to even remote rural areas all across the nation. It was a recognition of the tremendous social and economic benefits people could reap from electricity. The participants at the conference—labor leaders, representatives from farm organizations, national politicians, and public power advocates—drafted a bill for a nationwide public power system based on Thompson's ideas. The bill was endorsed by the Electrical Union, the American Federation of Labor, the National Grange, and the League for Industrial Democracy and introduced in Congress by Senator Norris and Republican Representative Oscar Keller of Minnesota. Although the effort for Thompson's Superpower failed, it set the seeds for similar ideas to emerge ten years later in the form of the 1935 Rural Electrification Administration and calls for a "national grid" that would echo into the 1960s and 1970s.

Private power's answer to the growing revelation of power company abuses and Thompson's Superpower proposal was a massive propaganda assault. The campaign was launched by a network coordinated by NELA and backed with funds that were unmatched in the history of American industry. The purpose was virtually to reshape how the public conceived of electricity and the private companies that delivered it.

Although proposals for such a massive advertising and propaganda effort had been circulating in NELA meetings since 1906, Samuel Insull was credited with finally putting it in place. He took most of the staff he had worked with on the state Council for Defense during World War I and made them into the Committee on Public Information (CPI) for his power fiefdom. Within the next few years, the NELA took over the organization and set up twenty-seven bureaus to cover all forty-eight states. To finance this elaborate system, each privately owned power company was assessed annual dues that were passed on as charges to consumers.

In the initial years a number of power company executives were doubtful of the new strategy. But M. H. Aylesworth, a former director of the NELA and an individual destined to become head of the National Broadcasting Company (NBC), assured delegates at a convention in Birmingham, Alabama, in 1924, "All the money being spent is worthwhile. Don't quit now. . . . Don't be afraid of the expense; the public pays the bills."

At the local level the industry's CPI made its way into the grade schools, high schools, colleges, libraries, and civic organizations. It also developed extensive influence with newspapers and radio stations. The all-embracing character of the propaganda campaign is suggested by Matthew Sloan, chairman of the industry's public relations campaign, who regretted that it was "perhaps impossible to make our public relations work stretch from cradle to grave." Sloan made a serious attempt to achieve that goal despite its ambitious proportions. Nearly a half million copies of pamphlets such as the two-page color pamphlet *OHM Queen* and another entitled *The Aladdins of Industry* were distributed to kindergarten and elementary school children to initiate the concept of private power companies as a beneficial force in the community. Similar propaganda work in the high schools and universities would have effects that would last for decades. Three fourths of the high schools in Illinois reportedly used specially prepared literature on the utility industry in their classrooms. The aim was to "fix the truth about the utilities before incorrect notions got there," according to William Mosher, author of *Electrical Utilities: The Crisis in Public Control,* published in 1929. The Illinois and Missouri committees surveyed textbooks and sent lists evaluating them as good, fair, or unfavorable to utility officials all across the country. Objections were made on the grounds that books favored municipal ownership or mentioned such things as a lack of competition in the industry, stock watering, or political corruption. In some states great success was reported in getting books removed. Joe Carmichael, director of the Iowa CPI, said the matter was taken up personally with school officials by local power company managers, and in nearly every instance books the private power companies objected to were removed from the Iowa classrooms.

The private power companies were less successful, however, in preventing the publication of textbooks containing "misinformation." The propaganda committees thought they had an understanding with leading publishers, who would submit the texts to the NELA. But according to Carl Jackson, chairman of the utilities' Committee on Cooperation with Educational Institutions (CCEI), the "people connected with higher institutions of education were very touchy on that subject and reports were not such as to encourage a very high handed attempt to directly control what should go into any textbook."

Some examples of the industry influencing the writing and editing of textbooks did emerge in federal investigations. Martin Glaeser's *The Outlines of Public Utility Economics,* a widely used text, was initially suggested by the CCEI and funded in part by the utility-financed Institute for Land Research and Public Utility Economics, an affiliate of Northwestern University. Glaeser testified that he was astonished to find that a set of proofs of the book had been turned over to the NELA before it was published. He admitted that he made some "corrections and changes of arguments" after receiving criticism from utility officials, but he defended the changes as "valid." When the book was published, it became a standard text for students of economics and utility engineering, helping shape the fundamental understanding of the industry. The utility industry reportedly helped distribute it. NELA also paid $3,000 to University of Toronto Professor James Mayor to write a book *Niagara in Politics* and another $10,000 to distribute it to schools, colleges, and newspapers. The book criticized Ontario Hydro, the successful Canadian public power system whose low rates were inspiring New York State officials to examine overcharges by private power companies. NELA also paid Professor A. E. Kennelly of Harvard and MIT, Professor C. F. Scott of Yale, and Professor Elihu Thomson $5,000 each for one year of study.

When the more benign propaganda and biased studies failed, the private power companies used political smear tactics. In the early 1920s, just after the Palmer Raids and "red" scares, in which thousands of people were arrested and deported for their political views, power industry propagandists centered their criticism on the notion of public power as "un-American" and tied to "bolshevik" ideas. In a small prelude of the communist scares of the 1950s, the Illinois committee went so far as to blacklist the United Society of Christian Endeavor, the American Farm Bureau Federation, the Religious Society of Friends, and the Women's Christian Temperance Union because of the organizations' support of public power systems. Apparently when the Teachers' Union of the AF of L objected to the committee's efforts to have certain books removed from schools and libraries, they also were blacklisted.

The private utility industry also relied on its army of employees to develop the myth that only private corporations should provide electricity. The New England bureau alone organized more than two hundred executives and employees as speakers on utility matters. According to the NELA, in 1924 the power industry's public speakers gave addresses to 6,000 audiences totaling 900,000 people. In 1925 speakers had addressed 10,000 audiences or a million and a half people. By 1926 the effort had nearly doubled, reaching 18,423 audiences and 2.5 million people, who were part of groups such as the Rotary Club, Kiwanis Club, Lions Club, school groups, real estate groups, or audiences gathered by the local Chamber of Commerce.

Newspapers were considered by power companies to be the most effective form of mythmaking. Most of the staffs of the state power company committees were headed by former newspapermen. J. B. Sheridan, the director of the Missouri CPI, wrote to a friend that he had spent "as much as $300 in three years entertaining editors . . . all of them are God's fools, grateful for the smallest and most insignificant service or courtesy." The industry spent approximately $28 million of its $33 million annual publicity budget on advertising in order to get favorable news stories in print. Articles and press releases from the Illinois committee alone went to nine hundred weekly newspapers. Many of them were reprinted as news items or editorials or feature articles. The Illinois committee reported 60,000 column inches of copy reprinted each year —the equivalent of four hundred newspaper pages of solid reading. In Georgia, the use of advertising money was credited with shutting out information on public power. Of 250 newspapers in the state that ran industry ads and copy, only four published anything from the "public ownership people."

The private power companies had direct ties with the radio industry as well. NBC was controlled by GE, Westinghouse, and the Radio Corporation of America. The president of NBC in 1928 was M. H. Aylesworth, the former publicity director for the NELA. At the regional level, power companies such as the Alabama Power Company owned radio stations. Many other electric companies ran weekly programs. Consolidated Edison of New York, for example, had a weekly radio hour. These programs offered the opportunity for more widespread propaganda. A forty-minute program aired through a forty-station hookup on October 27, 1927, celebrated the forty-eighth anniversary of Edison's invention of the incandescent lamp as Electric Night. J. F. Owens of the NELA gave the customary speech crediting America's achievements to the freedom of business from government interference and warned listeners that subversives were at work "marshalling their forces in an attempt to put the Government into the electric business."

When the power industry's influence over the media and the depth of this mythmaking was exposed, it was brazenly defended before the Federal Trade Commission. An investigation begun by the commission in 1928 filled more than eighty-four volumes with revelations on the shocking thoroughness of the power industry's propaganda, lobbying, and trading. Private power officials maintained it was "necessary to strike down misinformation and to keep dangerous political agitators in a strait jacket." They asserted that the primary danger threatening the industry was "the dissemination of false statements or erroneous information by misinformed or ambitious demagogues." The industry's leaders declared it imperative that the industry defend itself against an "unholy alliance of radicals."

There was more than just the propaganda and battles with power companies at the local level. In many ways the predictions of Tom Johnson and others on the ability of the power companies to corrupt institutions and destroy liberties were ringing true. At the same time this propaganda wave was washing over the nation, Carl Thompson, Judson King, and Senator George Norris witnessed the increasing concentration of political and economic control in the hands of the private companies. Senator Norris said, "The Power Trust already by the lavish expense of money controls numerous state commissions. It maintains an extensive lobby at Washington, headed by two former U.S. Senators. Other 'lame ducks' favorable to its design have appeared in certain federal commissions."

In one scandal related to this influence, Samuel Insull was called to testify before a special Senate committee. He freely admitted that he had donated $125,000 to the Senate campaign of Frank L. Smith, the former chairman of the Illinois Commerce Commission who had been responsible for regulating Insull's power companies in Illinois. At approximately the same time, newspapers in Pennsylvania disclosed that private utilities had made heavy financial contributions to the Senate campaign of William S. Vare in order to defeat a bid by Gifford Pinchot. These revelations garnered a rash of headlines and led the Senate to refuse seats to both Smith and Vare.

In the presidential campaign of 1932, Franklin Delano Roosevelt would denounce the power empire's information program as "a systematic, subtle, deliberate and unprincipled campaign of misinformation and propaganda, and if I may use the words—of lies and falsehoods." In 1928, four years before the industry became a presidential issue, the atmosphere of propaganda and corruption began to create huge clouds on the horizon for the private power companies.

What Insull and others had begun in expansion from regional to national control of power companies was reaching its logical extreme. According to one observer, Insull and others had gone wild in the rush of development of the holding companies in the mid- and late 1920s, flinging together "fantastic aggregates of geographically and socially unrelated systems scattered from hell to hallelujah." In addition to power companies, the holding companies also controlled strings of real estate companies, water companies, street and railroad ventures, and fuel and engineering firms. Their ventures ranged from the Philippines to central and southern Europe and South America.

Wall Street took an even bigger role in this final stage. Many of the sixteen large holding companies that controlled nearly all of the nation's power were now run by financial manipulators, who saw them as the perfect vehicles for reaping huge short-run speculative profits and building a pyramid of assets.

In the golden era of speculative stock trading, utility holding companies led the stock market and soaked small investors. One third of all corporate financing during the 1920s was issued by private power companies. Stock sales were based mainly on speculation over projections of steadily rising earning power. Even if an investor wanted to examine construction costs or the value of plants or assets, he or she could not do so, for corporate accounts did not disclose such costs and only in rare cases were appraisals available. In addition to sale of stock by brokers, the power companies sold stock directly through mail campaigns labeled "customer ownership" programs. Such sales were utilized not only to raise capital, but also to cultivate a large group of uncritical supporters. The NELA claimed that the 1.75 million stock and bond holders in private power companies constituted "true public ownership." In reality, however, a large number of the investors, as holders of bonds or preferred stock, had no voting rights. And common stockholders with voting rights were vastly outweighed by the moguls who held controlling interests in the power companies.

With a primary interest now focused on generating capital rather than electricity, the holding companies neutralized any advantages they had possessed over small isolated systems. The fact that their systems were now scattered all over the map disrupted cost-efficient integration. In their wars for control of territories and stock, they bought properties at exorbitant prices and raised rates for consumers. For example, competition with the Power Corporation of New York forced Associated Gas and Electric (AG&E) to pay between three and eight times the book value of several properties in upstate New York. In another scheme, Howard C. Hopson, the head of AG&E and a rival of Insull's, exchanged ownership of a block of stock thirty-seven times in thirty days among subsidiaries, in order to inflate the value several times over.

The string of nefarious dealings was endless. As one contemporary Wall Street analyst described it, the holding company directors were "wizards of financial chicanery," skilled at "double-shuffling, honey-fuggling, horn-swoggling and skullduggery." Needless to say, with Wall Street analysts unable to understand or follow the labyrinthine dealings of the holding companies, state regulatory agencies were hopelessly lost. There seemed to be no way to check the influence and power of these giant companies. The earlier inequities in regulation became even more exaggerated. Senator Norris said state regulation "can no more contest with the giant octopus than a fly could interfere with the onward march of an elephant." In a case brought by the City of New York in 1925 for lower rates, Consolidated Edison spent over $4 million in accounting, engineering, and legal fees. Most states at that time were spending less than $50,000 a year on regulating electric companies.

In 1928, William Mosher wrote that it was apparent utility commissions did not live up to original expectations. "Ratemaking in 1928, despite a score of years of effort on the part of regulatory bodies, is almost as hopelessly muddled, indefinite and unscientific as it has ever been." The results could be seen in rates for private power companies, which despite their advantage in size ran about seven cents per kilowatt hour. The average price of the small public systems by comparison was about five cents.

The profits of the service companies in the holding company systems also showed evidence of the enormous padding of rates. Electric Management and Engineering Corporation, for example, which served Insull's Middle West Utilities (MWU) group, collected $4.5 million in less than five years on which it made a profit of $2.8 million or 171 percent. AG&E between 1924 and 1927 collected $9.9 million in construction fees, on which it made a profit of $6.5 million or 193 percent. All of these inflated costs were passed through to consumers. The reality behind the myth of private power companies was that they had become milking machines, used to drain consumers and small investors in an effort to continue helter-skelter expansion.

Rivalry for control of the power industry heightened in the late 1920s as investment bankers sought to keep expanding their territory. The House of Morgan tried twice to take over the entire power industry. The efforts failed, but in establishing the United Corporation in 1929, Morgan interests quickly became the dominant force in the industry. Subsidiaries of the United Corporation produced 20 percent of the total electricity generated, GE subsidiaries accounted for another 13.6 percent for a total of 33 percent of the power generation under Morgan control. Insull had 10.3 percent of the rapidly expanding market. Electric service was delivered to two thirds of all the homes in the nation by 1928.

Amid the widespread rivalry and the atmosphere of scandalous abuse of consumers and investors, Ivy Lee—the "father of corporate public relations" who had been an inspiration to Insull—staged one of the greatest public relations gimmicks of his lifetime, the Golden Jubilee of Light. To celebrate the fiftieth anniversary of Edison's development of the incandescent bulb on October 21, 1879, Lee organized festivities in

Detroit, Michigan, attended by President Hoover, Henry Ford (who had been chief engineer at Detroit Edison [DELP] in his younger days), and other luminaries. President Hoover gave a speech broadcast by radio, and salutations to Edison were elicited from government leaders and scientists from all over the world.

The thrust of the event, of course, was to boost the scandal-ridden image of the private power companies by making them synonymous with Edison (note how many have incorporated his name or the name "public service"). J. F. Owens, president of NELA, would later write:

> I hazard the thought that the progress of the world in the last 50 years is the progress of electricity. As to when the electrical industry as an industry for the service of light, heat, and power began is a matter of argument. It is enough to say that the first real central electric station had its birth in September 1882, approximately 50 years ago. Its birth, or I may say coming into life for useful, practical work, occurred when the Pearl Street Station in New York was opened. To the ordinary mind and to the public generally at that time it meant very little. But to those ardent souls with venturesome spirits and with the strength of their convictions, who caused it to be built, it meant much. It meant for the American people later on and for all people in civilized countries a new era. It meant something that has become not only essential, but an actual necessity in our daily life and work.

Yet it was a statement by Albert Einstein that went to the heart of the event: "The great creators of technics, among whom you are one of the most successful, have put mankind into a perfectly new situation, to which it has not yet adapted." A week later, memories of the celebration were eclipsed by the collapse of the stock market, caused in no small part by speculation on huge volumes of power company stocks.

In the roar of the stock market just before the 1929 crash, Insull, Morgan, and others continued to draw in new investors through direct mail campaigns and aggressive sales. The new investors paid ten and fifteen times the price at which Insull had taken his shares. No one seemed to care. Commonwealth Edison went from $202 in January to $450 in August; MWU skyrocketed from $169 to $529 in the same period. In the fifty days ending August 23, Insull's securities appreciated at "round the clock" rates of $7,000 a minute.

In 1929, Insull was at the zenith of his power. His personal fortune had increased from $5 million in 1927 to $150 million in two short years. Besides a palatial house on Chicago's Gold Coast, he maintained a $3.5

million farm spread over 4,300 acres at Libertyville, Illinois. It was a "company town" in the most feudal sense. At this vast estate, over the preceding years, villagers were said to have built homes on Insull real estate, given birth to children in an Insull hospital and sent them to an Insull school, used Insull light, cooked with Insull gas, driven on an Insull road, and saved in an Insull bank.

The panic and stock market crash of October and November 1929 marked the beginning of Insull's slide from power. Confident that the ensuing Depression would be no longer or no more severe than those he had weathered in the past, Insull continued to expand his operations. His salesmen were still selling stock to the little investor with the pitch that it was more secure than government bonds. And his investment brokers continued to buy and sell power company stock in an effort to keep prices up.

As the nation's economy and the stock market careened ever downward in 1931, Insull and those around him borrowed frantically, putting up stock as collateral and stretching personal credit to the limit in order to keep the empire afloat. It wouldn't work.

Other events occurred as well that seemed to mark the end of an era. In the early morning hours of October 18, 1931, nearly two years after the Golden Jubilee of Light, Thomas Edison died. Charles Brush had passed away on June 15, 1929, while the planning for the Jubilee was going on. Although the NELA Bulletin made no mention of Brush's death, Edison was eulogized as the father of the electric industry. Nothing was said of how he was driven from the industry by Morgan in 1892. And in an effort to keep using Edison's image, the NELA—which had been widely discredited for its role in lobbying and propaganda campaigns—was dissolved. On the same day, January 12, 1933, the organization's officers created the Edison Electric Institute. Although acknowledging that the new organization was an "answer to demands by public utility leaders . . . to divest itself of all semblance of propaganda activities," the NELA executive committee installed themselves as the new officers, the "educational" materials of NELA were taken over by EEI, and offices were maintained at the same location on Lexington Avenue in New York.

As the era closed, nothing, not even cloaking their operations with Edison's aura, could forestall the disaster that was brewing.

3. The Turbulent Thirties

> [W]here a community, a city or a county or district is not satisfied with the service rendered or the rates charged by the private utility, it has the undeniable right as one of the functions of government to set up . . . its own governmentally owned and operated service.
>
> *Franklin Delano Roosevelt*
> *Portland, Oregon*
> *1932*

The day of doom for Samuel Insull arrived Friday, April 8, 1932. It began on an optimistic note, despite threatening rain clouds and a temperature hovering in the low forties. He arrived at the offices of an old friend, Owen D. Young, who served as chairman of General Electric and head of the New York Federal Reserve Bank. Insull was eager to discuss the status of some of his loans. Although his fiefdom had begun to teeter with the stock market crash of 1929, Insull had kept it afloat by stretching his personal credit to the limit, borrowing to pay interest and putting up his shares in companies as collateral. In February, Young had used his influence to gain an agreement from New York banks that they would not call Insull's loans due or ask for additional security until August. And just the day before Insull had tentatively arranged with representatives of a British investment house that had financed many of his earlier operations and an officer of Sun Life Insurance Company of Canada to lend him money to cover a $10 million note coming due on June 1. He had no idea he would be told to surrender control of his sprawling empire that very afternoon.

During the course of his discussion with Young and representatives of Continental Illinois and First National of Chicago, five men from Morgan's New York banks arrived. Insull was asked to leave the room. After an hour or so passed, Young appeared at the door and told Insull that the $10 million loan was not going to go through and that no one would put up money to cover the June 1 debt. Insull was described by one witness to have gone pale. He was faced with the largest bankruptcy in history. "I wish my life on earth had already come," he was heard to mutter. But his wish was not to be fulfilled. Investigation, flight, indictment, refuge in Greece, capture off the coast of Turkey, and trial in Chicago were yet to come. To millions of Americans, Insull became the symbol of the excesses and corruption of the huge corporations.

Behind the scenes it had been the Morgan organization that had taken advantage of Insull's weakness and broken his fiefdom. Insull's crash and that of several other utility fiefdoms was inevitable. Their soft spot was a dependence on operating company dividends and the sale of preferred stock to finance their operations. The stock market crash pulled these props from under them. The Depression drastically reduced flow of money from the sale of new stock and cut operating profits in half. As a result the holding companies had turned increasingly to the banks, piling up an intolerable burden of loans in order to finance their operations and continue expanding their territories.

Insull's two fatal mistakes had come in the rivalry for control with Morgan interests. In 1928, Insull's younger brother Martin, the president of Middle West Utilities (MWU), purchased a holding company system operating in fourteen eastern states. This resulted in a prolonged political battle in Maine over repeal of the Fernald Act, which prohibited the export of power to other states. Because of this law and Morgan's success in buying up neighboring utilities, Insull's companies were encircled and unable to sell more than a fraction of the power they were capable of producing. In response, Insull gave financial support to the power-consuming textile and paper industries in the state. This venture resulted in heavy losses, a serious blunder at a time when Insull could not afford it.

At the same time he began to fear takeover attempts from New York bankers. In late 1928 and early 1929 he reorganized his family's holdings into two new holding companies, Insull Utility Investment Corporation (IUIC) and the Corporation Security Corporation (CSC) of Chicago. These were placed at the top of the pyramid above MWU with each corporation owning a controlling part of the other.

Still Insull was afraid. In 1930 he made his final mistake by agreeing to purchase the stock holdings of Cyrus Eaton, a Cleveland industrialist who had been quietly buying up 160,000 shares of Commonwealth Edison, People's Gas and Light, and Public Service Company. Believing that Eaton was representing powerful New York banks, Insull agreed to buy back the shares at a price of $56 million. Already strained for funds by the New England venture, Insull turned in a moment of desperation to those he saw as friends in charge of New York banks. As collateral for their loans, he was forced to turn over some of the stock of his three key operating companies.

As long as the price of utility stocks, particularly those of the Insull group, remained strong, there was no cause for worry. Once the value of the collateral for the loans began to slip, however, Insull's empire would fold like a house of cards. Fortunately, prices remained amazingly stable until September 1931. When news arrived that England had gone off the gold standard, however, panic broke out on the New York Stock Exchange. During the week of September 8, the stock of IUIC, CSC, Commonwealth Edison, and MWU dropped by $150 million, and in the weeks that followed they plummeted. As the market declined, the holding companies had to put up increasing amounts of securities as collateral against their bank loans. By mid-December the well ran dry—every nickel of the combined portfolios of the two investment trusts was in the hands of bank creditors. During the next few months, Insull pushed his borrowing beyond safe limits, transferring money from one place to another in a desperate effort to shore up his companies. But the value of the stock continued to decline, and the bankers found themselves overburdened with Insull's notes and in debt to Wall Street.

The two banks in New York, National City and Chase National, which Insull believed could be counted on to remain neutral if not friendly, were faced with their own crises. Behind the scenes, these banks were also caught in the web Morgan was spinning. President Charles Mitchell of National City was indebted to Morgan for $10 million, and Albert Wiggin of Chase National, another of Insull's old friends, was evidently induced to participate in Morgan's scheme to take over Insull by a promise of participation in future financing of the power companies. Even before he had started for the meeting at Young's office on that gray day, Insull's fate was sealed and in the hands of Morgan interests.

It was ironic that bankers raised the ax to Insull the very day that nearly a hundred state and federal regulators attended a round-table conference on public utility regulation a few blocks away at the Hotel Pennsylvania. They called for federal incorporation of holding companies with adequate control over security issues, uniform systems of accounting, and full disclosure of their financial operations. James C. Bonbright, professor of finance at Columbia University and secretary of the Power Authority of New York, told the assembled delegates that the greatest structural weakness of the utility industry "is the superstructure of the holding company and of the so-called investment company, which in turn control that holding company." The holding companies had originally performed a useful service in consolidating the small competing utility companies, he said, but during the past decade the situation had drastically changed. Irrational consolidation, overcapitalization, and uncontrolled service charges were cited as the three reasons for the critical change that turned the holding company from "a public benefactor to a social menace."

The shock generated by the collapse of Insull's empire could be felt almost everywhere. Middle West Utilities, which was placed in receivership on April 16, controlled 239 operating companies, twenty-four holding companies, and thirteen miscellaneous subsidiaries serving 4,471 communities in thirty states and Canada. With the failure of the two top companies and MWU, all that remained was Insull's three major operating companies—People's Gas, Public Service, and Commonwealth Edison. For a time Sam Insull continued to direct these companies, but it was not long before the banks forced him to resign.

On Monday, June 6, 1932, the official end arrived. Insull spent most of the day signing resignations. He had been a director of eighty-five companies, chairman of sixty-five, and president of eleven. His son, Samuel, Jr., remained on to help the new management run the three operating companies. Insull stayed for a week in Chicago. He was broke and owed millions of dollars more than any assets he possessed. It was rumored in the street that he rode in a bulletproof limousine and was guarded night and day by four detectives. Although thousands of investors were wiped out when Insull's holding companies failed in April, there was apparently little personal hostility toward him. As one Insull executive put it, "If this had been an isolated failure, we'd have had to have every office manned with machine guns, but as it is, for most of the small losers, it's only one of many blows, and they seem to accept it fatalistically." The situation would change drastically as allegations of fraud emerged in the wake of Insull's flight to Europe and the Middle East.

Despite the lofty view from Insull executives of a fatalistic acceptance of the empire's collapse, the effect on investors was devastating. Insull had been in the forefront of the private utility industry's campaign to place more than $11 billion worth of its overpriced (watered) securities in the hands of investors, Insull's share being about one fourth of the total issues offered to the public in the six years preceding the panic in 1929. MWU alone had 73,447 common stockholders while industrywide the number was set between 3 million and 5 million. By 1934, the defaults on nineteen Insull properties represented nearly $200 million in obligations. The stock market losses were said to defy calculation, with estimates ranging from $500 million to $2 billion.

But these figures alone do not really tell the whole story. Hundreds of thousands of working men and women all over the country had poured their savings into utility stocks in the hopes of getting rich or at least securing a nest egg for their old age. A Chicago banker told the House Interstate and Foreign Commerce Committee that the "collapse of these companies has hurt our city more than did the great fire. . . . One day I stood and watched those holding securities and obligations of these companies coming in and filing them. They were average people—clerks and school teachers from Chicago, small shopkeepers from Illinois, farmers from Wisconsin—what they brought in, of course, was worth nothing." This scene was only part of the suffering that unfolded.

The damage also spread beyond these individual investors. A good portion of the capital Insull and other utility operators had raised in the 1920s came from banks and insurance companies. These institutions had invested hundreds of millions of depositors' dollars in utility securities. And given the interdependence of the utility industry, the banks, and the investors, it was no surprise that Insull's collapse would force calls on a multitude of loans, triggering tragedy for other utility empires as well.

When Central Public Service went into bankruptcy eight months after Insull, thousands of investors were left holding stock with little value beyond the price of the paper it was printed on. Stock that had sold for $80 now kicked around for as little as twenty-five cents per share. Henry Doherty's Cities Service half-million investors fared little better. The company suspended dividends shortly after the Insull disaster, and its stock fell from $68.12 to $1.25 at the bottom of the market in 1932. Associated Gas and Electric, controlled by Morgan interests, suffered a similar fate. As many as a quarter of a million investors, many of whom had purchased shares through sales campaigns reminiscent of the liberty bond drives of World War I, watched as the company suspended dividends on its A stock, and prices fell from a market peak of $72.16 in 1929 to a low of 62 cents per share in 1932.

In 1935 the FTC would report that more than ninety electric and gas companies had fallen into receivership. A private study of the power industry compiled in the spring of 1934 revealed that $600 million of debt in the form of bonds, notes, and debentures had been written off; four fifths of these defaults had occurred in the holding and investment companies.

Despite these failures and widespread scandal, the power industry stood up comparatively well during the Depression. With its monopoly status protected by regulation, it proved to be the nation's only substantially depression-proof major industry. At the height of the Depression in 1933, revenues were off only 11 percent. Gross agricultural income by comparison had fallen by 58 percent in three years, and manufacturing income declined 46 percent. Total power production dropped by about one third between 1929 and 1935. Much of the decline in industrial power usage was made up by increased sales to the 20 million households, which eventually provided one third of the electric industry's revenues.

The power industry seemed to survive on its momentum alone. Residential power usage increased 20 percent between 1929 and 1933 as tens of thousands of households purchased electric refrigerators, electric stoves, washing machines, irons, and other laborsaving devices for the first time. The use of electricity for air conditioning also added to the demand. Although air conditioning had been in limited use in manufacturing establishments for some time, it was not until the end of the 1920s that its use became more widespread. By that time it was starting to be used in auditoriums, restaurants, office buildings, trains, and stores. The revenue pouring in from the growth in residential use saved most of the operating companies. Philadelphia Electric Company for example, retained 90 percent of its 1929 boom year load and in 1932 turned over $18 million in dividends to its parent company, United Gas and Improvement.

But the momentum was not enough to carry the industry clear of the disaster surrounding Insull and the other holding companies. The growing need for electricity, and the way it had become an essential part of the American lifestyle, carried the industry directly into the political storm clouds on the horizon.

□ □ □

The 1932 presidential election was the first time in our nation's history that electric power stood out as a dominating issue. On one side was President Herbert Hoover assuring the nation that all was well and good times were just around the corner, the position maintained by the power empire. On the other side New York Governor Franklin D. Roosevelt hammered at those who had mismanaged and manipulated the economy and the electric power industry in particular. Convinced that the power industry was rife with corruption and that greater public control was essential, he proposed sweeping changes. The support for his position was reflected in a statement signed by thirty-seven Congressional leaders from both parties:

> We regard the power question in its economic, financial, industrial and social aspects as one of the most important issues before the American

> people in this campaign of 1932. . . . Its political significance cannot be overestimated and must challenge the attention of those interested in any progressive movement or measure. The reason is plain. The combined utility and banking interests, headed by the Power Trust, have the most powerful and widely organized political machine ever known in our history. This machine cooperates with other reactionary economic, industrial and financial groups. It is strenuously working to control the nomination of candidates for the Presidency and the Congress of both dominant political parties.

The statement was written prior to the Democratic National Convention in Chicago, at which power industry supporters initially blocked Roosevelt's nomination. According to political analysts of the day, what the power empire wanted was sympathetic candidates from both parties. The strategy to stop Roosevelt ripped wide open at the convention.

Among the aspirants for the top slot on the Democratic ticket could be found the names of Al Smith, who had run in 1928, Governor Albert Ritchie of Maryland, and Texas' favorite son, John Nance Garner, the Speaker of the House of Representatives. Although Roosevelt was at the front of the field, it was thought possible to stop him by undermining the two-thirds majority needed for nomination. If the balloting could be deadlocked, the way would be open for a dark horse candidate—a man with few or no delegates in the early ballots. John J. Raskob, a business executive in the Du Pont hierarchy who served as the Democratic Party's National Chairman, mobilized the national party organization against Roosevelt's nomination. The Raskob group, Roosevelt believed, was keeping Owen D. Young, the president of General Electric, on reserve. Young was seen as a consummate businessman who would negotiate the nation out of the Depression. It was later reported, however, that Young was made an offer and declined to run. The next choice of the power brokers was an Ohio attorney, Newton D. Baker

Baker's ties with the power industry went back to Tom Johnson's historic fight for public power in Cleveland. As a young man Baker had supported progressive causes, first serving as Cleveland city solicitor and then as mayor following Tom Johnson's retirement. At that time he had been a strong advocate of public power. He had later served as a distinguished Secretary of War and supporter of the League of Nations. By 1932, however, there seemed to be truth in the charge that Baker had changed his political stripes and was "a candidate in reserve for the plutocrats."

The record showed that he had long ceased to be progressive on economic issues. Baker was now one of the most prosperous lawyers in the nation, and his clients included Electric Bond and Share and the Van Sweringen brothers—the Cleveland financiers who controlled a $5 billion empire in railroads through the use of holding companies. He was also the chief counsel for power interests in suits challenging the constitutionality of extending federal control of water power on nonnavigable streams and rivers. And in 1924 he had written the foreword to a pamphlet urging that the Cleveland city power system be taken over by Cleveland Electric Illuminating, GE's subsidiary.

When the Democratic delegates met in Chicago to cast their ballots on July 1, they were showered with telegrams urging them to vote for Baker. Judson King claimed that the delegates he interviewed knew what was going on. "This is a power fight at bottom," they observed and speculated that "the stunt must have cost the Power Trust thousands of dollars."

For the entire first day of the convention and into the afternoon of the second, the balloting was deadlocked. Roosevelt had taken a lead on the first ballot, but remained 100 votes short of the two-thirds majority needed. Subsequent roll calls showed only modest gains. Finally on the fourth roll call, California swung its support to Roosevelt, with delegation chairman William McAdoo, a Garner supporter, striding to the platform to announce that California had come to Chicago not to deadlock a convention but to nominate a president. With California's support, Roosevelt's nomination was secure. Baker's candidacy and the strategy of the power empire for a dark horse victory hardly got out of the starting gate.

The efforts to block Roosevelt's nomination signaled a deepening of the struggle over electric power. Power industry executives were well aware of what Roosevelt had in mind. In his actions as governor of New York, he had shown himself to be a tireless opponent of private power conglomerates. His concern for cheap electric power dated back to his service in the New York State senate twenty years before when he had advocated harnessing the high tides of Passamaquoddy Bay in Maine for hydroelectric use. In 1927 he also advocated storing water on the high reaches of the tributary rivers to solve the Mississippi flood problem and told presidential candidate and New York Governor Al Smith that it could be done with a program "to develop hydroelectric power for the benefit of the people of the United States."

After being elected governor of New York in 1928, Roosevelt waged a three-year fight against private power interests over development of the St. Lawrence River. Roosevelt believed that the St. Lawrence, like other great water power resources of the nation, should not be given away to individuals or private companies. Across the river in Canada, the Ontario public power system was providing electricity at a fraction of the rates charged by private companies on the American side. The Hydro Electric Power Commission of Ontario had been established in June 1906, following battles with private companies similar to those that had occurred in the United States. Its success was a model for Roosevelt. He believed that a publicly owned dam and power house should be constructed as a yardstick to measure the private power companies' cost of producing and transmitting electricity. It was an idea for regulating the private companies that would provide the basis for Roosevelt's federal power proposals as president.

What he encountered in New York were Republican leaders in the legislature who had close ties to the Morgan interests (United Corporation through its subsidiary, Niagara Hudson Power Company, controlled 80 percent of the water power resources in New York). They had the bill for public development of the St. Lawrence killed in committee, preventing a general floor fight. An investigation would later show State Senator Warren Thayer, head of the power committee, held interests in Morgan power companies and kept up correspondence with Morgan representatives regarding legislation.

Roosevelt subsequently sent his personal associate, Louis Howe, around the state and the Province of Ontario to collect comparative data on electric bills from private and public power systems. What Howe found was that even within the state in 1929 prices had an absurdly wide range. A family using 250 kilowatt hours of electricity in Manhattan paid $17.50 per month, in Albany $19.50, in Buffalo $7.80. A family using the same amount of power in Dunkirk, which had a city-owned power plant, paid only $6.93. In Ontario the monthly bill from the regional public power system was $2.79 for 250 kilowatts.

Members of the legislature and the general public were shocked by the figures. In 1930, the legislature approved Roosevelt's proposal for a St. Lawrence Commission. In January 1931, the commission recommended that a state public power authority be established and a dam built across the St. Lawrence. Although the dam would be held up for a quarter century while American and Canadian authorities haggled over the terms of its use, the creation of the Power Authority of the State of New York was a huge step forward. Roosevelt appointed to this board men who were noted for their interest in the public welfare: Frank P. Walsh, Morris L. Cooke, James C. Bonbright, and Leland Olds. All were staunch opponents of the utility moguls and would become key figures in the New Deal's utility fights. The seed Roosevelt planted here would later flower into the Tennessee Valley Authority, the Bonneville Power Administration, and a massive rural program to utilize the nation's rivers to provide cheap power to remote farms and growing industries.

Roosevelt's stand on the power question, more than any other part of his public record, attracted Republican insurgents such as Senator George Norris and progressive Democratic senators such as Burton Wheeler of Montana and Clarence Dill of Washington. Sharing Roosevelt's conviction that public power could offer a solution to the abuses of the power industry, the bipartisan group of congressmen who supported him vigorously in his preconvention campaign would also support him in his New Deal legislative battles.

President Hoover's position was diametrically opposite to Roosevelt's. Hoover did not see any need to reform the utility industry in order to protect either consumers or the investing public. While serving as Secretary of Commerce under President Coolidge, Hoover had spoken out against federal regulation of the industry. He contended the power moguls were men of vision who were moved by the spirit of public service and that state regulatory commissions were providing all the protection consumers could possibly desire. "The majority of men who dominate and control the electric utilities," Hoover said, "belong to a new school of public understanding as to the responsibilities of big business to the people." His address, entitled, "Why the Public Interest Requires State Rather Than Federal Regulation of Electric Utilities," was given before the thirty-seventh annual convention of railroad and utility commissioners in October 1925 and was reprinted and distributed by the National Electric Light Association (NELA). Hoover was also opposed to the federal government taking any part in power production. In 1931, he vetoed Senator George Norris' proposal to authorize federal operation of the power plant at Muscle Shoals left partially built by the government during World War I. Hoover regarded this proposal as "pure socialism" and "the negation of the ideals upon which our civilization had been based." Under pressure from Roosevelt and public anger at power companies, however, Hoover was forced to moderate his stance.

In an atmosphere in which the Insull scandal pervaded the nation, it was inevitable that the enormous financial disaster would be picked up as a campaign issue. On September 15, 1932, two months after the Democratic convention and three months after Insull's collapse, the front page headlines in Chicago caught the attention of politicians all across the country. John Swanson, state attorney for Cook County, Illinois, was launching an investigation into the scandals surrounding Insull's dealings. Roosevelt tackled the issue the following week when he delivered a long attack on Insull before an audience of 8,000 in Portland, Oregon. The Portland Speech was the most thorough exposition of Roosevelt's positions on electric power and one which makers of New Deal policy would refer back to repeatedly during the tough decade ahead.

"The Insull failure," Roosevelt declared, "has opened our eyes. It shows . . . that the development of these fraudulent monstrosities was such as to compel ultimate ruin; that practices had been indulged in that suggested the old days of railroad wildcatting; that private manipulation had outsmarted the slow-moving power of government." The investors, who had placed over $1.5 billion in the "Insull Monstrosity," Roosevelt said,

> did not realize then, as they do now, that the methods used in building these holding companies were ultimately contrary to every sound public policy. They did not realize that there had been arbitrary write-ups of assets, inflation of vast capital accounts. They did not realize that excessive prices had been paid for property acquired. They did not realize that the expense of financing had been capitalized. They did not realize that payments of dividends had been made out of capital. They did not realize that . . . subsidiaries had been milked and milked to keep alive the weaker sisters of the great chain. They did not realize that there had been borrowing and lending between the component parts of the whole. They did not realize that all these conditions misstated terrific overcharges for service by these corporations.

For all of this, Roosevelt said, "The public has paid and has paid dearly and is now beginning to understand the need for reform after having been fleeced out of millions of dollars."

He also charged that state public service commissions had failed to live up to their high purpose of protecting consumers and the industry from its own machinations. "The Public Service Commission is not a mere judicial body, acting solely as an umpire between complaining consumers or complaining investors on the one hand and the great utility systems on the other," he said. "The regulatory commission . . . must be a tribune of the people, putting its engineering, accounting, and legal resources into the breach for the purpose of getting the facts and doing justice to both consumers and investors in public utilities."

Roosevelt also endorsed the idea of government ownership and operation of utilities when necessary to obtain service and fair rates. A community dissatisfied with the service rendered or the rates charged by the private utility had "the undeniable right as one of the functions of government to set up . . . its own governmentally owned and operated service." Expanding the St. Lawrence idea, he proposed four great power developments in the United States: the St. Lawrence project in the Northeast, Muscle Shoals in the Southeast, a Boulder dam project in the Southwest, and a Columbia River project in the Northwest. Once in place, he believed, the projects would strengthen local public power systems and "will be forever a national yardstick to prevent extortion against the public and encourage wider use of electricity."

The power issue in Roosevelt's view was multidimensional in nature —addressing not only electric power rates for homes and businesses, but also flood control, irrigation, and the preservation of water resources. He knew the fight that this would entail and told his Portland audience, "Judge me by the enemies I have made. Judge me by the selfish principles of these utility leaders who have talked of radicalism while they were selling watered stock to the people and using our schools to deceive the coming generation."

Democratic candidates in virtually every state where Insull's companies had operated echoed Roosevelt during the closing weeks of the campaign. Amid the din, the Hoover Administration announced the launching of a Justice Department investigation of Insull's last desperate attempts to save his fiefdom. On October 4, the Cook County grand jury, convened to consider Swanson's charges, returned indictments against Samuel and Martin Insull for embezzlement and larceny. Martin Insull was rumored to be in Canada. Samuel Insull had fled the country a week after resigning from his companies in June. He had first taken refuge in Paris, then Milan, and finally Athens. Hoover's attempts to have him seized and extradited on charges of using the mails to defraud, embezzlement, and violation of the bankruptcy act proved fruitless. It would be a year and a half later that Insull would be kidnapped from a tramp steamer off the coast of Turkey and turned over to American authorities. The atmosphere in which he would sail as a prisoner into New York was ironically different from the one that had greeted him when he first arrived to serve as Edison's secretary fifty years before. He became the scapegoat for the abuses that were rampant throughout the power industry. And the private power industry he had helped create would be on the defensive from a new surge in public power organization.

□ □ □

Roosevelt's election in 1932 marked a turning point in the historic battle between public and private interests for control of power resources and territory. In eight years, Roosevelt and the New Deal would transform the agenda of American politics. While in 1932 politicians were absorbed to an astounding degree with such questions as prohibition, World War I debts, and law enforcement, by 1936 they were debating social security, valley power authorites, holding company regulation, and public housing. The creation of the Tennessee Valley Authority, the Bonneville Power Administration, and the Rural Electrification Administration would set an environment for new growth in publicly owned power systems. And the development of rural electric cooperatives serving areas previously untouched by private or public power would bring a revolution to rural life in the form of milking machines, refrigerators, electric lights, and other appliances.

The private companies fought to head off the growth of public power as they had so effectively done in the early 1920s. At every step of the way in state legislatures, in the courts, and at the ballot box they argued public power was inefficient, and they raised the age-old charge that the Tennessee Valley project and other public power efforts represented "creeping socialism."

The opening salvo in what would be a decade-long battle for control of the industry was fired when Roosevelt called upon Congress during the first one hundred days of his administration to approve the Tennessee Valley Authority Act. The spark for the new authority was the long unresolved question of what should be done with the partially constructed government nitrate plant at Muscle Shoals on the Tennessee River in northern Alabama. In 1917, as part of the government's munitions works for World War I, President Woodrow Wilson had picked Muscle Shoals as the site for two nitrate plants to be powered by a dam and two coal-fired generating plants. At the war's end, construction was discontinued when Congress declined to authorize additional funding.

For thirteen years disputes had raged over what to do with the site. The debate pitted defenders of public power against the private power interests and their allies in the Coolidge and Hoover administrations. George Norris, as chairman of the Senate Agricultural Committee, led the fight against the attempts of private interest to take over the project. Norris, who claimed he never knew how the issue came to be dropped in his lap, saw the struggle as an "irreconcilable conflict between those who believed the natural resources of the United States can be best developed by private capital and enterprise, and those who believed that in certain activities . . . the federal government itself can perform the most necessary tasks in the spirit of unselfishness for the greatest good to the greatest number."

Few persons of influence were prepared to declare, like Norris, for complete government ownership of the electric power industry. But allied with him were defenders of the public power tradition and supporters of conservation and unified river development. Among his backers could be found the names of several prominent organizations including the AF of L, National Farmers' Union, the National League of Women Voters, the Methodist Federation for Social Service, the Public Ownership League, and the National Popular Government League. His opponents, who spoke with one voice for private development of Muscle Shoals, included the National and Investment Bankers Association, the U.S. Chamber of Commerce, the National Electric Light Association (NELA), the National Manufacturers Association, and several large insurance companies.

Through his unyielding efforts Norris won enough votes twice, in 1928 and 1931, to have legislation for public development passed by Congress. But both Coolidge and Hoover vetoed the bills. His success had been greater in stopping private development. The most threatening bid had come from Henry Ford, who offered to purchase the unfinished project in 1921. Ford planned to develop Muscle Shoals into one of the great industrial centers of the nation. His highly publicized 1927 tour of the area with the aged and approving Thomas Edison encouraged a real estate boom. The area surrounding Muscle Shoals was incorporated, streets were laid out, and the property subdivided. Special trains ran from New York filled with prospective land purchasers, while airplanes carried passengers into the sky to view the sweep of the country where a city rivaling New York was to rise when the project was turned over to Ford for development.

Before the speculative bubble burst, tens of thousands of dollars would be siphoned from the pockets of honest men and women, who invested their life savings in the mistaken belief that they were getting in on the ground floor of a new American Wonderland. In the end, Henry Ford's dream of a new southern metropolis died in the Senate, a victim not of "Wall Street," as Ford had claimed, but a coalition that cast both Norris and his private power company enemies together in an effort to stop a huge new trust from taking over.

The mighty Wilson dam remained largely idle and the nearby nitrate plants silent and deserted through the 1920s. With Roosevelt's election, the deadlock was broken. The new president told Congress on April 10, 1933, that Muscle Shoals represented only a small part of the potential usefulness of the Tennessee River. If envisioned in its entirety, Roosevelt told them that development would transcend electric power to include flood control, soil conservation, afforestation, retirement of marginal farmland, and diversification of industry. To provide a unified direction, he proposed that Congress create "a corporation clothed with the power of Government but possessed of the flexibility and initiative of a private enterprise"—the Tennessee Valley Authority.

The sweeping proposal was just what the private power companies had feared. It was met with staunch opposition, based on arguments that the power wasn't needed, and an ideological debate that the power business should be left to private enterprise. The power company opposition was led by Wendell L. Willkie, the forty-one-year-old president of the massive Commonwealth and Southern holding company (C&S) controlled by Morgan interests. C&S was one of the largest holding companies in the nation, with thirteen subsidiaries operating in eleven states, spanning an area from Florida to Michigan. The holding company was responsible for the power of more than 5 million customers, and its combined assets were valued at $1 billion. The Tennessee Valley Power Authority proposal was seen as a cancer in its heart.

In an appearance before the House Military Affairs Committee, Willkie cleverly told the chairman that his company was not opposed to the government's plan to provide cheap power at Muscle Shoals as long as it was sold to the power companies at economy rates and distributed through privately owned transmission lines. What he strongly objected to was the building of government transmission lines. Such a project without transmission lines, however, would benefit only the private companies. Willkie warned that, if Congress approved the building of transmission lines as part of the new power authority, C&S would lose $400 million in value of its securities. He stated that the Tennessee Valley was more than adequately served by C&S systems. His cast of supporters would amplify that claim.

Preston Arkwright, the president of Georgia Power, told the committee, "I do not think we need additional lines, and no power is needed to serve that territory." E. A. Yates, a vice-president for C&S, would state flatly, "I can see no market value whatever for this power. It is a market being served completely and fully and at reasonable rates by these companies. If the plan was carried through, there would be additional excess capacity." Their statements before Congress would prove hollow. They later offered to purchase all of the power authority's output, and a new market grew up for steadily increasing amounts of electricity.

When the TVA Act was considered by the full House, Republican congressmen predictably raised the well-worn cries of "bolshevism" and "communism." Representative Joe Martin, a Republican from Massachusetts, for example, declared that the TVA was "patterned clearly after one of the Soviet dreams." Charles Eaton, a representative from New Jersey, commented, "This bill, and every bill like it, is simply an attempt to graft onto the American system the Russian idea." While there was agreement on the dire need to revitalize agriculture in the Tennessee Valley, the version of the bill passed by the House placed limitations on the federal government's ability to build dams and transmission lines. Norris, who took charge of the fight in the Senate, was able to push the bill through without any restrictions. Roosevelt intervened in favor of the Senate bill when it was taken up by a joint conference committee, insuring that the House would go along with the original measure.

After so many years of trying to make use of the Muscle Shoals facility, Norris and his staff were ecstatic. "It was a glorious fight right up to the very end," Norris' secretary wrote to a friend, "with President Roosevelt standing firmly behind Senator Norris in every particular." Although the legislative battle was finally over when Roosevelt signed the TVA Act on May 18, 1933, the new governing board of three directors would face stiff resistance from the private power companies as they attempted to carry out the law. Roosevelt chose the three directors carefully. Arthur Morgan, president of Antioch College and a professional engineer, had done extensive work on flood control. He noted in one interview that his beliefs had been shaped by reading Bellamy's book, *Looking Backward,* when he was fifteen years old. Harcourt Morgan, president of the University of Tennessee, was a man who had devoted his life to agricultural improvement and would oversee the valley authority's farm program. The third director was a thirty-four-year-old Chicago lawyer, David Lilienthal, who had served on the Wisconsin Public Service Commission and was noted for his progressive thinking. Lilienthal would lead the power program and eventually make an immense impact on the future of the power industry.

The main opposition to the valley authority continued to come from the Morgan-controlled C&S company. Right at the start Willkie sat down with David Lilienthal in a private meeting at the Cosmos Club in Washington. Willkie stunned the young lawyer by telling him that the New Deal was only a passing fancy. The project depended on Congress for its funds and he could find the federal government's generosity gone at any moment. He suggested that Lilienthal make the TVA independent of Congress by selling all of its power for about a half million dollars a year to C&S.

It was a curious proposal from a man who had insisted the power was not needed in the region. When they left the meeting, Lilienthal was shaken by Willkie's cocksureness.

While legal challenges to the power authority were being filed and fought in court, the Tennessee project pushed ahead at the grass roots. The most immediate problem was developing a market for the electricity. Most of the rural homes in the region had no electricity. And few transmission systems spanned those areas. In 1932, there were only sixteen municipal power systems in Tennessee serving 14,000 customers. Most of these systems purchased power from C&S, and in view of the rumors Willkie had spread about the TVA being a short-lived dream, many were reluctant to sever their dependency on the private company.

In February 1934, Lilienthal finally made an agreement with the town of Tupelo, Mississippi, to begin providing wholesale electricity. This first contract enhanced the power authority's position and gave at least one indication that the project could go forward despite Willkie's influential opposition. In June, the power authority marked another milestone. The Alcorn County Electric Power Association began operation as the first rural electric cooperative in the valley. Organized in the back room of a furniture store in Corinth, Mississippi (population 6,220), the cooperative delivered the power authority's electricity to farmers and townspeople at half the price they had paid for electricity from Mississippi Power Company, C&S's local subsidiary. The nonprofit farmers' cooperative was able to pay back approximately half its debt for purchase of Mississippi Power Company's lines in the first year of operation. This promising start would eventually open the way for an entirely new public power program for rural cooperatives, but first the mounting trouble with C&S and other private power conglomerates would have to be stopped.

Lilienthal and the other two TVA directors insisted that each community had the undeniable right to own and operate its own electric plant as "one of the measures which the people may properly take to protect themselves against unreasonable rates." This right could be realized by acquiring existing plants, setting up new plants, or even building competing plants, as circumstances dictated. After the rapid drop in power bills for the first of the municipal systems contracting with the TVA, takeover fights began to break out all across the valley. Between 1933 and 1938 seventy-two elections were held in cities and towns on the question of establishing a publicly owned power system and contracting for TVA power. In these fights private companies did everything they could to defeat the takeovers, even resorting to illegal tactics.

In Chattanooga, home turf of the Tennessee Electric Power Company (TEPC), an organization known as the Citizens and Taxpayers Association formed to campaign against a takeover referendum. A significant number of the group's members were employees of the company. C&S contributed $20,000 of the $22,000 budgeted for the organization's activities. The TEPC also chipped in by purchasing two vacant lots and deeding them to 120 nonresident employees so that they could vote on election day. The "citizen" group also supported company opposition to the Chattanooga *News,* which had editorialized in support of the takeover. A new newspaper, the *Free Press,* was transformed from a grocery store chain circular and subsidized by the power company. On election day, in addition to handing out reams of company propaganda, the group also distributed free liquor. The Tennessee Public Utility Commission would later find TEPC guilty of 1,917 counts of using "unlawful devices" in the election. Despite these tactics, the proponents won a smashing victory. Voters in Chattanooga, like their counterparts in sixty-two of the seventy-two other communities where elections were held, overwhelmingly opted for public power.

Assisting in the financing of the public takeovers was the Public Works Administration (PWA). This agency, established in June 1933 to provide employment relief, had authority to provide funds for the construction of municipal electric systems. Its loans gave public officials in Chattanooga, Memphis, Knoxville, and many smaller communities the leverage they needed in the difficult and prolonged process of negotiating a buy-out once a takeover vote had passed. Where the private companies resisted selling their transmission systems, the towns could threaten to build duplicate systems and attract the private company's customers with lower prices.

In Norris' home state of Nebraska, loans from the PWA and the Reconstruction Finance Corporation, another government program, provided money for establishing three public power districts in 1933 and 1934. As in Tennessee, hundreds of thousands of acres of farmland in Nebraska were wiped out by flooding and drought during the "dust bowl days" of the Depression. Similar to the TVA, the public power districts sought to build dams for irrigation, flood control, and electric power. While Nebraska's proposal would be attacked in local and national press and political forums as an idea supported by "parlor pinks" and "bolsheviks," it would prove enormously successful in saving farmland and producing electricity. Eventually all of Nebraska would receive public power, and no private power companies would be left operating in the state.

The loans and grants were also a boon to new public power systems in the Northwest, where Roosevelt's promises for damming the huge

Columbia River system were being fulfilled. By 1935 the PWA had received applications for 495 public power projects. Allotments totaling $50 million were awarded to cities and towns, and in many others the threat of a new or competing municipal system forced private companies to lower their rates. Suits mounted by private power companies against the program were launched all over the country as they were against the TVA. By the fall of 1938, the private power companies challenged the constitutionality of TVA in fifty-seven different actions. Progress of both programs was delayed, but their mandates were finally upheld by the Supreme Court in 1938.

After the court decision, the private companies quickly negotiated favorable settlements. In September 1938, the TVA and the city of Knoxville jointly took over the Tennessee Public Service Company and its 34,000 customers for $7.9 million. Three months later, the Kentucky and Tennessee Light and Power Company transferred to the TVA and several cities property serving 10,000 customers in northwestern Tennessee. And in April 1939, Willkie, acting on behalf of C&S, agreed to sell all of TEPC's property for $78.6 million, thus extending the power authority's electricity to another 142,000 customers in Nashville, Chattanooga, and other cities throughout the state. By the start of World War II, the TVA would provide electricity to seventy-six municipal plants and thirty-eight rural cooperatives. It would also come to manage some twenty-one dams and build new locks and shipping canals, increasing river traffic tenfold on the Tennessee River. And its other programs would protect lowlands from flooding, provide incentive for industrial development, and give the nation the first "installment buying" program, allowing farmers to pay on time for equipment they purchased.

While the TVA was a cornerstone in Roosevelt's power program, it not only provided a yardstick to measure the rates of the private companies, but it also gave some measure of the difficulty local communities would have in trying to exercise their option to choose a public rather than a privately controlled electric system. Before other power programs would be launched, Roosevelt went on the offensive against the massive holding companies.

□ □ □

As the nation labored in the depths of the Depression and David Lilienthal worked to get the TVA off the ground, Samuel Insull was brought back to New York to stand trial for embezzlement, violation of bankruptcy laws, and mail fraud. His arrival caused a sensation. The public and the media regarded him as the personification of big business gone bad, the kind of wheeler-dealer the public held responsible for the stock market crash and the flood of misery that swept the nation. In essence, Insull was the scapegoat for the abuses widely practiced by the investment community and power moguls. For eight weeks one of the most costly trials in American history was staged in Chicago. Seventy-four-year-old Insull was presented as a man caught in the excesses of building and maintaining his sprawling industrial conglomerate and serving society's electrical needs. He was said to have won over the jury with a touching presentation of how he had risen from the position of office boy to head one of the nation's largest power conglomerates. His abuses and those of other leaders of the power empire were lost under the shadow of Insull's personal story. While he was publicly condemned, the jury acquitted him of all charges.

Eventually, however, executives of the private power companies and their principal stockholders would have to pay part of the penalty. Long-smoldering anger had been kindled by the revelations of the FTC investigations on lobbying and propaganda. Details of the Insull machinations and the underhanded tactics used to stop public power in the Tennessee Valley, Nebraska, and the Northwest fueled that anger further. Roosevelt and his Brain Trusters seized the opportunity to try to halt the common abuses heaped on consumers and investors and break up the holding companies. The fight over legislation they introduced would be the most fiercely fought battle of the New Deal, and it would touch off one of the most intense and scandalous lobbying campaigns in American history.

Benjamin V. Cohen, general counsel for the National Power Policy Committee, drafted the bill that Burton K. Wheeler of Montana introduced in the Senate and Sam Rayburn of Texas brought to the House on February 6, 1935. Turning the tables on the private power conglomerates, the power policy committee charged that the holding companies had gone beyond serving normal business functions and economic needs and developed a form of "private socialism" in which their combined corporate interests practiced grave abuses on American investors and consumers. The solution was to eliminate the holding companies that served "no demonstrably useful and necessary purpose" and to place the remaining holding companies under strict federal regulation. Those operating far-flung companies would be forced to divest and break down to a single contiguous system. Subsidiary businesses such as real estate ventures and railroads would be minimized to functional purposes. Service companies and mining affiliates that had previously bloated their charges would be forced to supply services at cost. And most importantly, ownership of controlling shares by Wall Street firms would be prohibited.

The broad view was that the Public Utility Holding Company (PUHC) bill aimed at reforming the practices of Wall Street as much as those of the power companies.

The uproar from the financial community and the power industry was unsurpassed. They attacked the bill as unconstitutional, unnecessary, and dangerous for the industry and the nation. The opposition was coordinated by the Edison Electric Institute (EEI), betraying claims that the new organization would not continue the lobbying and propaganda of the defunct NELA. Working through the Committee of Public Utility Executives, the EEI relied on the most sophisticated techniques of advertising then available to persuade community leaders the administration's policy was opposed to the American Way.

T. J. Ross, a partner in Ivy Lee and Company, the public relations firm that had organized the propaganda campaigns in the 1920s and was responsible for the Golden Jubilee of Light in 1929, was given the task of rallying investors, bankers, and business groups against the bill. Ross attempted to create what historian Phillip Funigiello has termed a "conspiracy neurosis." His goal was to persuade nonutility companies that the New Deal was a plot to destroy business and the free enterprise system. Insurance executives, for example, were informed that company assets would be seized and used to fund holding company debts. Bankers were told banking institutions would be included in the definition of holding companies and become subject to its "destructive measures." And retailers, distributors, and manufacturers were given information that they would be classified as holding companies too, making their investments vulnerable to provisions of the bill.

Investors were also organized against the bill. Blyth and Company, a utility underwriting firm, sent a letter to more than 3,000 investment bankers across the country, warning that the bill would "disastrously affect the value of all public utility securities including those of operating companies and seriously retard business recovery." Accompanying the letter was a do-it-yourself kit describing the most effective techniques that utility salesmen could use to mobilize their clients in opposition. The Bank of New York sent reprints of an editorial to its clients written by the conservative journalist David Laurence, entitled "New Dictatorship." In a cover letter, the president of the bank, J. C. Traphagen, warned that the bill would reduce the purchasing power of thousands of securities owners. Numerous front organizations, such as the American Federation of Utility Investors, were formed to try to put a public face on private industry's campaign. The "federation" was later found to be supported primarily by producing pamphlets for the power industry and through donations of brokerage firms and large investors. The New York *Journal of Commerce* described the organization in 1934 as a "movement of stockholders inspired by utility companies and launched in Chicago in order to gain the appearance of independence from Wall Street."

The pressure on Congress was unequaled. According to the Scripps-Howard press, there were more individual lobbyists working on the issue in Washington than there were members of Congress. Avalanches of mail and telegrams buried the desks of legislators. The freshman senator from Missouri, Harry S. Truman, received more than 30,000 letters and telegrams. Most were later discovered to be phony. In one incident, a Senate investigating committee headed by Hugo Black discovered that associates of Howard Hopson had forged names from a Pennsylvania city directory. Shortly after the fraudulent telegrams were sent, a Hopson employee told the Western Union office manager in Warren, Pennsylvania, that it would be a good idea if "somebody threw a barrel of kerosene in the cellar" of the telegraph office. The manager told the investigating committee he later found the charred remains of the copies of telegrams in a basement stove.

A whispering campaign was also discovered to have been initiated to impugn the motives of Roosevelt and other New Dealers. Early in the summer, as the debate was headed toward its peak, an advertising man was said to have suggested to S. Z. Mitchell, the president of Electric Bond and Share (EBASCO), that suspicion could be spread that the New Dealers and Roosevelt were sick or insane. Thomas McCarter, the president of EEI, gave currency to this rumor when he told a conference of bankers that the president had an "obsession on the subject" of holding companies. "It is a condition of mind that even many of his closest associates in Washington do not understand." The rumor was floating in other circles too. By mid-July, *Time* magazine reported that its Washington correspondents were plagued by inquiries concerning the president's mental health.

A long series of hearings on the proposed holding company legislation was held from January through April 1935. The focus of much of the attention was on what was known as the "death sentence" clause, which called for the compulsory dissolution of holding company systems after 1940. A compromise later authorized the federal Securities and Exchange Commission to disband any holding company that could not prove its usefulness.

When the bill came out of committee, Hugh Magill, head of the American Federation of Utility Investors, told the press that the legislation "constitutes the most sinister peril to our Democratic form of Government that has ever threatened our people." Despite the unsurpassed opposition and utility spending estimated at $1.5 million, the bill moved onto the floor in both chambers.

In the Senate, proponents of the measure carried the day with the "death sentence" intact, by only one vote. In the House, however, the bill was passed but without the "death sentence." In the end, Roosevelt had to accept a compromise allowing for federal regulation of the holding company systems that remained in existence. It was a weaker law than he had envisioned, but passage of the PUHC Act of 1935, nevertheless, represented a significant victory over the private power empire and would force a partial restructuring of power corporations. It made control from Wall Street more difficult and loosened, for a time at least, the threat of "corporate socialism" that the power empire had posed to the nation's economic and political structure. It would also allow further expansion of public power systems.

In the midst of the fight over the PUHC Act of 1935, Roosevelt issued an executive order establishing the Rural Electrification Administration (REA). Like the TVA, the rural electrification plan was not a new one. George Norris, Judson King, and others had long advocated bringing electricity to remote farms and villages. Roosevelt's personal interest had been perked in 1924 while undergoing treatment at Warm Springs, Georgia, for the aftereffects of polio. When he received his first power bill at Warm Springs, he found the local rates were four times as high as those for his home at Hyde Park, New York. This led him to consider the issue carefully when it was brought before him.

While Roosevelt didn't commit himself to a rural power program during his election campaign, the success of the first TVA cooperative at Corinth, Mississippi, set an encouraging precedent. Morris Cooke, who had served as chairman of the Giant Power Board during Gifford Pinchot's governorship in Pennsylvania and as Roosevelt's Power Authority commissioner in New York in 1931, urged the president to expand the program. To bring the system into being Cooke also concentrated much of his attention on Harold Ickes, the Secretary of the Interior.

On several occasions he presented Ickes with a proposal for a joint program with private companies. The time was ripe, Cooke argued, because the private power companies would work to help. In one meeting, Ickes listened to him in silence and responded, "I'll have nothing to do with the sons-of-bitches." Cooke was not prepared for the reply, but sensing an opportunity asked, "Then will you consider a plan wholly under control of public authority?" Ickes responded by asking if public electrification was feasible. Cooke told him yes. "Then shoot," Ickes answered.

Cooke also lined up his support in Congress. In October 1934, while serving as head of the Mississippi Valley Committee, Cooke repeated his pleas for federal leadership in rural electrification. Judson King and the National Popular Government League also urged Roosevelt to direct the government to finance such a program. In the Senate, George Norris, fresh from his Tennessee Valley victory, called for prompt federal action. Norris' boyhood memory of chores done in the flickering light of a coal-oil lamp "in the mud and cold rains of the fall and the snow and icy winds of winter" had left him with a passion to bring electricity to the farms of America. "I could close my eyes," he said, "and recall the innumerable scenes of the harvest and the unending, punishing tasks performed by hundreds of thousands of women . . . growing old prematurely; dying before their time." The American Farm Bureau Federation and the National Grange also passed resolutions at their annual meetings in 1934 urging federal action. The Farm Bureau wanted the Farm Credit Administration to finance cooperatives and mutual light and power associations. The Grange wanted a system that would deliver power to people "at the lowest cost possible."

Although pressure for a rural electrification program continued to mount, it was the gloomy economic situation that finally persuaded New Dealers to throw their support behind the program. With 9 million men unemployed and whole families lacking food, clothing, and shelter, federal relief on a massive scale was needed. The vision of electrifying over a million farms, of providing rural dwellers with the conveniences of urban living, and of stemming the flight from the farm to the already overburdened cities finally struck a responsive chord.

In his annual address to Congress in 1935, Roosevelt urged rural electrification as part of a new public works program. In April, Congress passed an Emergency Relief Act, authorizing nearly $5 billion in expenditures for public works projects, including $100 million for construction of rural electric lines. As the vicious fight over the PUHC Act moved into a final few months, Roosevelt issued an executive order establishing the REA. Morris Cooke, still carrying the ideas from Pinchot's Giant Power Board in Pennsylvania, was named the first administrator of the program.

In 1935 only 11 percent of the nation's farms had electricity, and the costs for a private company to bring it in were enormous. Rural customers were usually required by private companies to pay between $2,000 and $3,000 per mile for line extensions. In addition, families could expect to pay between ten and twelve cents per kilowatt hour, and in some regions as high as twenty-five and forty cents. At a time when wheat sold for eighty cents a bushel and gross farm income averaged $1,800 a year, electricity was an undreamed-of luxury.

Initially, Cooke hoped that, by gaining the cooperation of private companies and providing them with low-interest government loans, there would be rapid development of rural power networks. The clash over the PUHC Act, however, had created a great deal of hostility. A committee of utility executives told Cooke in July that "there are very few farms requiring electricity for major farm operations that are not now served." Outraged, Cooke turned to the municipal systems for assistance. But they were still hemmed in by laws preventing them from extending their lines beyond city borders. Cooke also faced difficulty in the fact that 90 percent of the labor for the rural projects had to be hired from the relief rolls. Few skilled electrical workers could be found on the rolls. As a result of these problems, only a handful of farms received power in the first year of the program's operation.

In January 1936, George Norris and Sam Rayburn set out to correct the weaknesses in the project by taking it out of the relief program and making it a separate statutory agency. While the bill was bitterly contested in the back rooms of Congress, the power industry mounted no major public opposition, believing it would be "another New Deal flop." Industry executives couldn't accept the fact that farmers would be willing to go into debt for electricity or that they would be able to succeed, given their lack of legal and financial skills. The prevailing attitude was best expressed by a utility executive who said, "Let the farmers build electric cooperatives; then when they fail, we will buy them up at ten cents on the dollar."

With the new law, Cooke focused on forming rural cooperatives. Loans would still be awarded to private power companies for developing rural extensions, but preference was given to nonprofit farmer groups. As the program began to take off, private power companies were forced to embark on their long overdue rural projects. Many built "spite lines" through the center of proposed REA districts in order to claim that any other construction would be unnecessary duplication. In other areas, when it appeared certain that a cooperative was to be organized, they would "seize the cream" by rushing a construction crew in to build lines for the most lucrative customers, leaving the poorer ones without power.

These practices were evidently widespread. Harry Slattery, who served as the rural program's second administrator from 1939 to 1945, said that during the first five years approximately two hundred cooperatives were affected by the private companies' tactics. A total of 15,000 miles of line and 40,000 customers were reported lost by 192 of the co-ops, and eight of the new co-ops were wiped out entirely. Overall it was estimated that 100,000 consumers had been left in the dark because of the private companies' spite lines.

Other strategies were used by the private power companies as well. Squads of canvassers were sent out in automobiles, driving from farm to farm to discourage farmers and their wives from joining a cooperative. "Rural people are told a thousand and one cooked-up tales to bewilder and terrify them," Slattery said. "They are solemnly assured that when a farmer signs for membership in a Rural Electric Administration cooperative, he is, in effect, putting a mortgage on his property." Farmers were told that they would be individually liable for construction costs and for any damage that might arise out of the construction or operation of the system. And if the cooperative failed, they would lose their farms to the government.

There were also legal campaigns to stop the formation of cooperatives that mired farmers in time-consuming litigation. Such was the plight of seven hundred farmers who tried to organize a tricounty project in northwestern Massachusetts. As there were no laws permitting the formation of a consumer power cooperative in Massachusetts, the farmers sought to organize as a private membership-controlled company. This required state approval, and the private companies successfully intervened to have the application denied. The setback discouraged farmers elsewhere in the state from forming cooperatives.

But despite these and other tactics, the REA was a huge success. Farm women especially, who stood to have an enormous burden lifted from them, played a special role in organizing the cooperatives and sign-up campaigns. By 1941 there were 8,000 borrowers in the program. Among them were the names of 741 new cooperatives, fifty-four municipal or public utility district agencies, and twenty-five private companies. By 1944, 43 percent of the nation's farms had electrical service. In every region, it renewed the sense of miracle that had been spreading in urban areas for over fifty years.

Joe Brawner, a lineman for the first rural electric cooperative in Mississippi, said, "It was the greatest thing in the whole business. We wired the houses, brought out the appliances, put in the meter with the family crowded around waiting. When the first switch was turned on, they literally cried and shouted with joy." A Texas rancher who had witnessed the oil booms "where enthusiasm was a fever heat, where it cost two dollars a plate for ham and eggs" said he had "never seen such enthusiasm as is now shown in Deaf Smith and Castro counties over the coming of REA electricity." In one rural community families staged a mock funeral for a kerosene lantern to signify the end of life without modern conveniences.

Families accustomed to a life of toil and drudgery no longer needed to spend up to ten hours a week pumping water and carrying it from a nearby well to the kitchen. Nor did they need to scrub clothes in an outdoor tub or kettle filled with hot water heated by a fire. Running water and indoor bathrooms connected to electric pumps reduced a host of medical problems associated with contaminated water. And electric refrigerators reduced food spoilage and the number of cases of dysentery, undulant fever, and gastrointestinal diseases. Electrical equipment also reduced a farmer's work load. Although gasoline tractors had begun to reduce the amount of labor and time required to do field work, farmers still relied on muscle power for a variety of mundane tasks such as drawing water, feeding livestock, milking cows, and making repairs, which could be performed much more efficiently with electrical machinery.

Given the hardships that rural people faced in desolate fields or dark forests, it is no wonder that they responded so enthusiastically to the march of poles across the countryside. A circus atmosphere often prevailed when the REA farm demonstrations were set up in fields outside towns. Sometimes as many as 5,000 people showed up from neighboring counties to see the wonders of electricity. Inside the big tent, local women's organizations prepared hot meals for visitors, using electricity to demonstrate that life in the kitchen could be simplified. Meanwhile, in the field, farm men stood in awe as hammer mills, corn shellers, ensilage cutters, hay dryers, and pig brooders were demonstrated.

The broad success of the program was attributed in large part to policies that enabled farmers to obtain low-cost service without having to pay for construction costs of individual line extensions. Under its "area coverage" policy, the REA constructed a backbone electrical distribution system adequate for providing service to everyone in an area. The determination for a loan was based on the resources of the entire system, rather than individual line extensions figured in the profit estimates of private companies. For farmers who couldn't afford to make a financial commitment of an extra $1.50 a month to finance the line, the REA staff devised a plan of self-help under which a cooperative's members could work off a portion of the expense through labor on line construction and home wiring. The cost of line construction was also drastically reduced. Before REA, private power companies were spending $1,500 to $3,000 per mile on construction of power lines. By 1940, the co-ops were building lines for as little as $500 per mile, without sacrificing reliability.

The success of the program fed not only into the growth of TVA, but also into the establishment of the other federal projects Roosevelt envisioned as "yardsticks" against which the private companies would be measured. Federally assisted projects like the financing of the Santee-Cooper hydro project in South Carolina and the Loop River Public Power District in Nebraska and the Bonneville Project Act of 1937, which fulfilled Roosevelt's promise to the Northwest to harness the power of the massive Columbia River, gave a broad base to the yardstick.

With public power agencies given first choice or preference rights to this federal power, these projects also boosted a resurgence of the municipal power movement. In addition to the 741 cooperatives, sixty new municipal power systems came into being between 1932 and 1937. Many of these were "public utility districts," which extended outside city limits to cover a county or several towns. California had been the first state to authorize these public agencies in 1913. Arizona, Nebraska, and Montana followed in 1917 and Michigan in 1927. By the end of World War II, eighteen states had authorized public utility districts. The most extensive efforts came in Oregon and Washington, especially after the Bonneville Project Act.

Despite the support of the federal government, private companies kept up their attacks to stop the resurgence of public power. A Federal Power Commission (FPC) report, issued in February 1941, summarized an investigation into the political activities of power companies in the Northwest. The report revealed that five private utility companies had spent more than $1 million in propaganda efforts to stop the formation of public utility districts and other public power systems. A primary strategy was the formation of citizen front groups acting on behalf of the private companies. Commissioner John Scott wrote that "by subterfuge these companies sought to pollute the political process of free choice at public elections." Funds obtained from their operating revenues had been lavishly expended, Scott said, to prevent people from obtaining electricity through publicly or cooperatively controlled organizations.

The struggles over who would control electricity ultimately led to disruption of policies and programs, which left the nation short of power for war production. In March of 1938, President Roosevelt, fearing the approach of war and well aware of the shortages created by the struggle between public and private interests before World War I, instructed the

FPC to cooperate with the War Department in surveying the nation's power capacity. The confidential report submitted four months later predicted that if wartime production demands were placed on the nation's electric systems, there would be critical and widespread shortages of power. The report recommended that the nation's power load be diversified as much as possible among existing plants and factories within fifteen critical war production centers—primarily located east of the Mississippi River and north of the state of Tennessee. While the nation would need only 1.1 million kilowatts to meet a peacetime demand in 1940, the report predicted a need of 5 million kilowatts for the fifteen centers if the country was to go to war.

During the summer of 1938, two rival plans emerged for assuring an adequate power supply. The first, supported by the FPC, brought back the plans for a Superpowerlike system—constructing an integrated network of high capacity transmission lines supplemented by the expansion of hydroelectric and steam plants. The other plan, backed by the Army and Navy, proposed to increase capacity by building new steam plants in each of the fifteen critical war production centers.

The FPC plan gradually emerged as the preferred approach, preserving the government's stake in the power field. The network would be constructed, maintained, and operated by the federal government as a national defense measure and operate as a common carrier during peacetime. The new hydroelectric plants would be undertaken as joint ventures by the federal and state governments and the new steam plants, financed by low-interest government loans, would be constructed by private power companies in coal-mining areas.

By the fall of 1938, a National Defense Power Committee was established to examine various proposals and draft enabling legislation. It was not long before the issue of public and private control disrupted the committee. The conflict was carried on largely by two men: Louis Johnson, the Assistant Secretary of War who chaired the committee, and Harold Ickes, Secretary of the Interior. Johnson, who had lobbied against the PUHC Act on behalf of the private power companies, proposed that private enterprise should underwrite the financing and construction of the transmission network with as little duplication as possible by the federal government. If Johnson succeeded, Ickes feared that the private companies would take advantage of the system in the event of a war emergency, placing its henchmen in key government power agencies and undermining the New Deal's successful rural and public power program. His fears were not without foundation. One private power official,

J. C. Damon, told William O. Douglas of the Securities and Exchange Commission that the PUHC Act was an obstacle that would have to be removed for private power's cooperation in the defense program.

As the international situation deteriorated in the spring of 1939, the National Defense Power Committee became paralyzed by the conflicts. The committee was reshuffled, but in the final analysis the United States entered the war short of power, and Ickes and the private power interests shared the blame for delaying implementation of a new national power program.

At the national level the struggle continued too. It came as no surprise to most observers that Roosevelt's opponent for the presidency in 1940 was none other than Wendell Willkie, president of the Morgan-controlled Commonwealth and Southern holding company system and the champion of the private power interests. This time it was not a "dark horse" race. Although Roosevelt was re-elected with 27,243,466 votes, Willkie shadowed him with 22,304,775.

The upheaval of the war in the next five years ultimately brought extensive expansion to both private and federal power projects. But beneath the acceleration of the industry and the social tumult, the question of who would control the power industry would become an even further-reaching issue at the end of the war.

4. *The Coming of Nuclear Power*

> I should point out that any reference to the "atomic industry" is perhaps not altogether correct. In fact, there is not today, nor is there going to be in the future, a separate atomic energy industry. . . . Rather atomic energy will encompass all the industrial activity of the nation. It will, in time, cut across our entire industrial and economic pattern. In time, also, each existing industry will become concerned with some phase or some particular aspect of atomic energy.
>
> *Paul F. Genachte, director of Atomic Energy Division of the Chase Manhattan Bank, 1956*

At 5:30 A.M. on July 16, 1945, when the world's first atomic blast blew devastating heat and light in a towering mushroom cloud over the desert and mountains near Alamogordo, New Mexico, the world was changed forever. For the electric power industry especially, there was to be a wave of "new fire" that would heighten the struggle between public and private interests and spark sweeping changes over the next four decades. Nuclear power promised an impact ten or a hundred times that of the first coming of electricity. The radioactive wastes of atomic energy alone contained hazards that would last thousands of years. And in the business world Wall Street executives calculated that atomic energy would eventually permeate every aspect of industry. Whoever controlled its development and use would hold the same cornerstone that the early electric power moguls had staked out at the center of the nation's economy.

Officials of both public and private power systems understood this. Roosevelt had temporarily sidetracked the domination of the electric power industry by powerful financiers, but atomic technology provided a new vehicle to restore their drive for corporate control of the nation's electricity systems. More than ever before, devious machinations in the back rooms of federal government overshadowed democratic processes. Ultimately, the struggle for private domination of the electric industry pushed past unanswered questions on atomic costs and safety, laying unimagined hazards before communities all across the country.

Coming on the heels of the Depression and the upheaval of World War II, the birth of atomic energy unleashed emotions that ranged between unknowing fear and blind hope. Somewhere in the middle was a fascination with unlimited abundance being produced by such a "god from a machine." It had a long build-up in the popular imagination of the nineteenth century.

Following the tremendous accomplishments of the industrial revolution and the initial experimentation with electricity, the late nineteenth century seemed to be a time when the most ancient secrets of nature were about to be harnessed for man's use. In January 1896 the German scientist Wilhelm Roentgen stirred the world by announcing the discovery of X rays. Although others had already constructed X-ray machines, Thomas Edison leaped into this new field with the same speed he had entered electricity. By May 1896 he exhibited an X-ray machine at the National Electric Light Association (NELA) in New York City. Other experimentation with X rays and radioactivity brought about more revelations. In 1909 Frederick Soddy, a British chemist and physicist, summarized recent experimentation in a book called *The Interpretation of Radium and the Structure of the Atom* and estimated that the amount of energy in a ton of uranium would "light London for a year."

Soddy's work inspired H. G. Wells's 1914 science fiction novel *The World Set Free.* With uncanny prophecy, Wells anticipated the dangers and benefits from atomic energy. Following a devastating World War in which atomic bombs destroyed major cities, Wells foresaw a utopian period he called the Efflorescence, in which use of atomic energy would be widespread. Forty years before the first experimental reactors were developed, he wrote, "It was in 1953 that the first Holsten-Roberts engine brought induced radio-activity into the sphere of industrial production, and its first general use was to replace the steam-engine in electrical generating stations." Although he incorrectly assumed that the atomic bomb would come later, Wells was off by only a year concerning electrical use of atomic energy. In December 1951, at an experimental reactor near Arco, Idaho, sixteen Atomic Energy Commission (AEC) staff members watched as steam produced by heat from the reactor turned a turbine and lighted four 200-watt bulbs. That light had the same dazzling effect on the scientists as the early demonstrations of arc lamps by Charles Brush had on viewers more than seventy years before.

Following the devastation of Nagasaki and Hiroshima, Congress and the scientific community were eager to prove the beneficial applications of this miraculous new energy. It promised an abundance that would eliminate the scars of the Depression and the horrors of war. In the months after the war, David Lilienthal, head of the Tennessee Valley Authority and someone destined to play a critical role in the formative years of the new technology, said people envisioned that "a revolutionary period based upon the peaceful use of atomic discoveries lay just ahead." According to Robert Bacher, a physicist who worked at Los Alamos and later served on the AEC, the mid-1940s was a time of great excitement among scientists. "Many technical people felt that the development of reactors could take place with great rapidity," he said. In popular media, too, visions of atomic-powered airplanes, trains, and cars, deserts made fertile, irradiated food preserved without refrigeration, medical isotopes, nuclear-powered electricity, and of course "the bomb" were widespread. In an atmosphere rich in foreboding and hope, what quickly evolved was a three-way wrestling match among scientists, the military, and Congress over who would control the development of nuclear power.

Even before the first test bomb had been exploded at Alamogordo there was a jockeying for position. Several months before the war's end, General Leslie Groves, head of the Manhattan Project, which developed the atomic bomb, had planners drafting legislation to secure military control of research and development of atomic energy. Scientists who were insiders to the process opposed military control. In early 1945, before the Alamogordo test, they had also drawn up a plan. It called for keeping nuclear research free of military and political interference. Vannevar Bush, a leading scientist with the Manhattan Project, proposed federally sponsored research overseen by a governing board controlled by civilian scientists nominated by the National Academy of Sciences.

In fall of 1945, just a month after the Hiroshima and Nagasaki blasts, when Congress returned from its summer recess, the debate burst into the open. On October 3, President Truman sent a message to Congress stressing the necessity for legislation on atomic energy. By December a bill crafted by Senator James O'Brien McMahon, a progressive freshman Democrat from Connecticut, came out of a special committee he chaired. Although the military would lose control, the new plan would allow continued military development for weapons parallel to research for peaceful uses. Technical control would be placed in the hands of appointed scientists, and Congress would maintain oversight through a new congressional committee. Because the British and the Russians were expected to develop atomic technology soon, President Truman pressed passage of the legislation and for formation of a compatible foreign policy aimed at control of atomic technology by a new international agency.

On March 5, 1946, a five-man commission headed by David Lilienthal and Dean Acheson of the State Department presented a report on the need to share information for research on peaceful uses of atomic energy and to outlaw the development and possession of nuclear bombs. Three months later, Bernard Baruch, an international financier and United States delegate to the United Nations, offered to that body the Acheson-Lilienthal proposal. Baruch said that the real problem with atomic energy was not one of physics but one of ethics. The Russians, who were still three years from exploding their first atomic bomb, also put forward a proposal.

It was a time for far-reaching decisions. While the United Nations headed into frustrating debate on international nuclear control, Congress pushed toward passage of domestic nuclear legislation before its summer recess. In early July, Lilienthal published an article that laid out the dilemma as one facing more than just official decision makers. He wrote:

> The dominant fact of our time, is the towering place of the machine, of applied science, in the lives of mankind. And the great issue of our time, with which the peoples of the whole world will be at grips day in and day out for the rest of our lives, is simply this: Are machines and science to be used to degrade man and destroy him, or to augment the dignity and nobility of humankind . . . ? From this issue no one who lives today can escape.

It was not just the threat of the bomb that Lilienthal was worried about. He also feared those who worshipped technology:

> Efficiency is their god, and the managerial elite are their high priests. They broadcast radio programs full of the romance of gadgets and fill the slick-paper magazines with odes to a chromium bathtub. Technology they seem to say is above good and evil. If the spirit of man balks, if being human increases the cost of production—well, then the man must be redesigned to fit the assembly line, not the assembly line revised for man.

Having seen what could be accomplished by programs such as TVA, Lilienthal knew firsthand the enormous power of technology. He believed that the decisions on how atomic power was to develop should be made by people in a democratic process, based on full knowledge of the risks and alternatives. Things such as the share of the national budget that would go to nuclear research, the adequacy of protection against health hazards from radioactive materials in the air and on the ground, and how rapidly atomic fuel might supplement coal and oil and hydropower as a source of electricity depended in his view "upon a sacred and inviolable process, the dissemination of knowledge."

Albert Einstein was urging a similar course of action. In an interview in *The New York Times Magazine* on June 23, 1946, he said:

> During the war many persons lost the habit of thinking for themselves, for many had simply to do as they were told. Today such a lack of independent thinking would be a great error, for there is much the average man can do about the danger facing him. . . . Current proposals should be discussed in the light of the basic facts. They should be discussed in every newspaper, in schools, in churches, in town meetings, in private conversations, and among neighbors. . . . Our representatives in New York, Paris and Moscow depend ultimately on decisions made in the village square. To the village square we must carry the facts of atomic energy. From there must come America's voice.

The Acheson-Lilienthal proposal was stalled in the United Nations, but the idea of an appointed civilian commission guiding atomic development in the United States stuck. In late July, just ten months after Truman's request for legislation and only a year after the first atomic bomb blast, Congress passed the Atomic Energy Act of 1946. Truman signed it on August 2. Out of the initial wrestling match won by the scientists and public-minded officials came a second struggle over whether the government or private companies would develop atomic energy. Of particular importance was the prospect for nuclear-powered electricity. The struggle for control of the electric power industry, which had pushed through hundreds of fights in cities and towns and raged through the federal government for more than a decade, now took on broad international proportions entangled in the secrecy and national security crises over atomic energy.

The Atomic Energy Act set up a five-person commission to oversee atomic development in cooperation with a joint committee represented by both houses of Congress. David Lilienthal was nominated to be the chairman of the commission, Truman's testimony to his success as a director of the TVA. Others included William Waymack, Pulitzer prize-winning editor (1938) of the Des Moines *Register Tribune* and one of the nation's most prominent journalists, Dr. Robert Bacher, a young physicist from Los Alamos who had worked on the Manhattan Project as well as the United Nations proposal, Sumner Pike, a Maine businessman and former member of the SEC, and Lewis Strauss, a reserve admiral in the Navy and conservative investment banker from the Wall Street firm of Kuhn Loeb. While Lilienthal would play an important role in the early struggle between private and public interests, it would be Strauss whose religious commitment to free enterprise would ultimately shape the course of accelerated nuclear development.

Backing the commission was a General Advisory Committee of scientists, led by Robert Oppenheimer and Enrico Fermi, who advocated basic research and cautious development of atomic technology. The commission was also influenced by researchers and scientists working for major corporations such as Westinghouse, Dow Chemical, and General Electric, who had been under contract to the government from the beginning of the Manhattan Project. Many of these individuals, oriented toward private development, felt restrained by the cautious attitudes of the General Advisory Committee.

The unsteady compromise over government-versus-private development of atomic energy was reflected in the AEC's basic mandate, which attempted to straddle the issue:

> Subject at all times to the paramount objective of assuring the common defense and security, the development and utilization of atomic energy shall, so far as practicable, be directed toward improving the public welfare, increasing the standard of living, strengthening free competition in private enterprise and promoting world peace.

Under a rationale that secrecy had to be the guiding force, the act was interpreted by the Democrats to give the federal government a complete monopoly over development and ownership of facilities in nuclear industry. This frightened free-enterprise advocates. Also included in the act was a clause giving public power systems a preference to electric power from any federally funded and built nuclear plant, similar to the clause for hydropower from federal dams. For private power company officials, it was as if the public power systems suddenly had access to a source of unlimited electricity. This tension set the stage for a long period of conflict.

Brien McMahon was appointed head of the first congressional Joint Committee on Atomic Energy on August 2, along with eight other senators, including Tom Connally of Texas and Harry Byrd of Virginia. Among the nine members from the House were Claire Booth Luce of Connecticut and Lyndon B. Johnson of Texas. It was the first time in history that a congressional committee had been given such sweeping powers over such a secretive force. Some writers have speculated that the committee's powers both to legislate and regulate were in violation of the Constitution. The Democrats who dominated the committee and President Truman, who had approved their vast authority, set themselves up for a huge surprise.

In the November 1946 elections, the Republicans took control of Congress. Not only was McMahon forced from the chairmanship of the committee, but the commissioners who had been appointed by President Truman, especially David Lilienthal, who was to chair the commission, faced rough going from those who saw him as the nation's gleaming symbol of public power. The confirmation hearings for the new commissioners went through thirty-two public sessions and six executive sessions in the process of interrogating fifty-five witnesses.

During the hearings the champions of private power companies in Congress attacked both Lilienthal and the interpretation of the act that left atomic development to the federal government. The Atomic Energy Act was charged with creating an "island of socialism." Senator Kenneth McKellar of Tennessee, who had supported proposals for the military to maintain control over atomic development, aimed outrageous questions at Lilienthal and others in an attempt to have his name withdrawn.

In questioning Wall Street financier Bernard Baruch, McKellar asked: "Knowing that you are not a communist, do you think we ought to select for this commission a man who in his organization at home has had in the last ten years somewhere between forty and fifty-five communists, and whose department has defended all communists in his setup down at the TVA? Do you think that at this time, with the world in the shape it is in now, with the United States having the questions confronting it that it has now, do you think it is the time to appoint those friendly to the communistic cause?"

"But I do not know to whom you refer," Baruch answered.

"I'm talking of Mr. Lilienthal," McKellar said.

Like many others, Baruch attempted to deflect McKellar's slur. "I would not say he is friendly to the communistic cause, Senator, from what I have seen of him," he replied.

When Lilienthal came before the committee, McKellar accused him directly of having "communist sympathies" and questioned him on the birthplace of his parents in Czechoslovakia. Lilienthal denied the charge, described where his parents were born, and pointed out that an investigation of TVA in 1940 by those looking for Communist Party members had turned up three suspected communists out of 18,200 employees.

Although Lilienthal would be confirmed as head of the AEC, McKellar and those worried that Lilienthal would limit the access of private corporations to atomic technology would continue their allegations of communist sympathies and public power conspiracies.

Shortly after the confirmation hearings, major industrialists pushed for a committee of industry representatives to be given security clearances and allowed to study applications of atomic energy. Formed in 1947, the Industrial Advisory Commission was led by James Parker, president of Detroit Edison. Secretary of the eight-man group was Walker Cisler, an engineer and consultant to the Manhattan Project and the future president of Detroit Edison (DELP). Cisler was a driving force in private industry's development of atomic power. At the close of the group's study, he would report that "secrecy" and other barriers could be reduced and that industrialists looked forward to "significant expansion of industrial participation" in the "not distant future." It was indeed ironic that while the general public would not be offered information regarding far-reaching choices and decisions about atomic energy, private companies would be allowed to shape these decisions. That irony would deepen as the political machinations produced an evasion of a key provision of the Atomic Energy Act and a gross betrayal of public trust.

The general climate for the struggle over nuclear power was also affected by other conflicts over the control of electricity. Nationally, there was the breakup of the holding company empires. This gave a burst of growth to local public power systems. By 1946, Nebraska was well on its way to being the only state in the nation to have all its electricity served by public power systems. Internationally, there was also a steady trend toward public power spurred by reconstruction after the war. In April 1946 France had nationalized 1,200 electric power companies, five groups of which had controlled 74 percent of the country's power. In Great Britain a bill to nationalize electricity was introduced in the House of Commons in December 1946. It passed Parliament and received royal assent in August 1947, combining 180 private companies and 360 public systems under a central British Electricity Authority.

Although two thirds of the electric systems in Britain had been publicly owned for decades, the heated debates on nationalization of the private companies had a very familiar ring. Part of the British nationalization act stated:

> The municipal undertakings regard the supply of electricity as a local public service. The companies regard it as a business enterprise. The local authori-

> ties base their claim on the fact that Parliament had throughout designated them as the ultimate owners of the industry. The companies, always conscious of their limited tenure, have repeatedly tried to show that the technical development of electricity, which requires undertakings covering wide areas, was best served by the strengthening of company organization.

While nationalization of the electric power industry in the United States was unimaginable, there was a potential for the spread of locally controlled power systems. It was clear that development of more federal power authorities similar to TVA, Bonneville Power Administration, and the Southwestern Power Administration established in 1943 would strengthen existing public systems and provide a strong base for voters to create more rural cooperatives and municipal power systems. Fearing this, in 1945 the private systems had set up the National Association of Electrical Companies "to whittle away at public power projects" and prevent additional "river valley power authorities" from being developed. In Congress, the dominant Republicans sought to isolate the public systems by curtailing construction of federal transmission lines. The House Appropriations Committee reported that it was "against the purpose of our form of government to appropriate Government funds for the construction of transmission lines, switchyards, substations and incidental facilities when private capital is prepared to provide them." According to the private industry plan, in order for municipal and cooperative systems to purchase power from the federal government, they would have to contract with the private companies to transmit it.

As part of their resurgence from the Roosevelt era, in 1947 the private power industry began what Philip Sporn, head of the American Gas and Electric Company, called "a five year, nine billion dollar expansion program—a program unprecedented in its history, or for that matter, in the history of any other industry in this country . . ." At the local level the private companies restarted their ancient strategy to stop the growth of the federal projects, which would feed new public systems. With the backing of a Republican Congress, the private power companies consciously sought to cut off funding and construction for federal transmission lines and generating plants.

One key fight erupted over a new coal-fired plant at New Johnsonville, Tennessee. It was to be the TVA's first nonhydro plant that was not related to Defense Department work. When Republicans in the House succeeded in blocking the New Johnsonville plant and other federal power projects, Truman called their actions "bad, foolish, reckless, irresponsible and a capitulation to special interest groups." During the 1948 presidential campaign, he spoke out against the Republicans in words that echoed Franklin Roosevelt. "The Republicans don't like to sell cheap public power," he told an audience in El Paso in September 1948, "because it means that the big power monopolies cannot get their rake-off at the expense of the public." He warned that if the Republicans maintained control of Congress they would next scrap the clause providing public and cooperative systems a preference to federally generated power. This would allow private companies to dominate federal power distribution. "Now you know what that means," Truman said; "that means that the . . . utilities lobby will have control . . . to turn this electric power over to the hijackers, so they can stick you with higher prices."

Truman's surprise victory and the Democrats' return to power in Congress in 1948 provided a temporary respite from the attack on the federal power program. With the coming of the Korean War, TVA would eventually receive funds to build six new coal-fired plants. For the nuclear program, it was a different story.

As head of the AEC and the shining symbol of public power's success with the TVA, Lilienthal was accused of slowing down development of nuclear power and keeping the government monopoly closed. In an address to the New England Council, a group of 3,000 businessmen and manufacturers, in November 1948, he explained his position and why public rather than private development of reactors was essential to national and world security. "As things now stand," he said, "an atomic furnace or an atomic power plant is virtually an atomic bomb arsenal." In order to prevent the proliferation of bombs, he said, "the usual industrial and business formula is not applicable, certainly not at this juncture in world events, and probably never will be completely. This undertaking is so close to issues of life and death for our country that the kind of development—the amount of manpower and funds, the direction and the pace of development—these must of necessity be determined not by private considerations but considerations of public security." He said the ownership and full responsibility for design and operation of reactors "cannot be delegated to private hands."

Under Lilienthal, GE would operate the Hanford military reactor program and Carbon and Chemical would run the Oak Ridge facility under a government contract system established during the war. But the participation of private industry remained largely limited to mining and milling of uranium and supply of equipment. Lilienthal's concern, as well as that of Oppenheimer and Fermi, was that development of atomic technology be thoroughly researched and tested. Those favoring more rapid commercial development continued political attacks. Lilienthal's critics claimed he was waiting for the time when government-owned reactors would sell power in the same way Bonneville and the TVA did, creating a vast new power base for municipal systems and rural co-ops, which were still largely dependent on private power systems for their electricity. The scenario of public nuclear systems emerging in Britain and France was the recurring nightmare of American private power companies and their financial allies.

For nearly a decade there had been a significant growth of public power going on. In 1942 the American Public Power Association had been formed out of a need to combine the interests of local public power systems. The organization's small staff monitored federal legislation out of its Washington office and sought to coordinate the policies and efforts of the 2,078 public power systems serving towns and cities across the country. That same year the National Association of Rural Electric Cooperatives was organized to coordinate the efforts of 803 rural systems. In the background of this consolidation of rural and public systems, a major fight was going on over "freezes" placed on co-op materials and projects. Private power companies, on the other hand, were being given adequate materials and a green light for expansion. Co-op managers charged that the War Production Board was staffed by utility executives who were taking advantage of the war to thwart the Rural Electrification Administration. Kansas Representative Thomas Winter made counterallegations that the cooperatives were promoting "socialization of electricity" and that the Rural Electrification Administration was hoarding copper. Congressional hearings over the charges of copper hoarding resolved the issue in favor of REA. But throughout the war years, allegations of communist conspiracies continued to surface.

Amid much controversy the Southwestern Power Administration, a marketing agency similar to Bonneville Power Administration in the Northwest (though less than one tenth its size), was formed in 1943 to serve Kansas, Missouri, Oklahoma, Arkansas, Louisiana, and Texas. In 1950 the Southeastern Power Administration would be formed to provide power from Army Corps of Engineers projects to ten eastern states. As these agencies strengthened local public power systems and rural co-ops, they sent a deepening scare through the private power industry, and gave the private companies an even greater desire to gain dominance over nuclear power.

In 1949 the private power industry conducted a study to determine the best way to arouse public sentiment against government power programs. Officials of the Electric Companies Advertising Program, a project formed in 1941 by the Edison Electric Institute, were shocked to learn that 63 percent of the people interviewed in a poll approved of TVA. The supporters weren't fuzzy thinkers but upstanding citizens who had substantial wealth and education. Test ads placed in *Time* magazine and *The Saturday Evening Post* revealed that emotional themes such as the spread of "socialism" produced the greatest response. The lesson to be learned they said was to "stress the fight against the socialist state more in the future."

Their interest in that lesson was intensified when a special Air Force unit picked up a cloud of radioactivity over the Pacific in early September 1949. It was evidence that the Russians had exploded their first atomic bomb. The world was stunned. Although Russian atomic power was anticipated, the blast had come sooner than expected. The haunting aura of an atomic confrontation that had lurked in the background of the Cold War now became the preoccupation of the new nuclear age. The whole atmosphere of atomic development shifted. Emotional support rose for development of a superbomb—a hydrogen bomb proposed by Edward Teller as the ultimate weapon. Lilienthal and Oppenheimer opposed the project and the dangerous escalation it could lead to. President Truman was caught in a fiery debate that divided scientists and the AEC. In late January, Truman decided to build the bomb. Faced with an overriding priority being placed on the project, David Lilienthal resigned as head of the AEC two weeks after Truman's decision.

□ □ □

On Lilienthal's departure in July 1950, Gordon Dean became the new chairman. Dean, an attorney from the antitrust division of the Justice Department and a friend of Senator McMahon's, had been named to the Atomic Energy Commission only a few months before the Russian atomic test. He provided the first major shift in the struggle for private control of nuclear power. His initial steps were taken under the overwhelming mandate for atomic weapons development and in the midst of communist witch hunts of the McCarthy era, which held suspect any scientist or government official who questioned the nation's atomic bomb program or the interests of private corporations that dominated the programs. Amid growing tension, Dean's steps pushed past serious questions of whether nuclear-powered electricity was needed or would ever be practical.

Fear of the Cold War escalating into an atomic confrontation was magnified by the North Korean crossing of the 38th parallel on June 24 and the opening of the Korean War. By October 1950 the AEC had launched a thirty-three-month crash program to build and demonstrate Edward Teller's H-bomb. The $3 billion program included massive expansion and huge new plants at Oak Ridge, Tennessee, Paducah, Kentucky, Hanford, Washington, and Portsmouth, Ohio. It represented one of the greatest federal construction projects in peacetime history. Although only a fraction of the budget was allocated for reactor research and development, the potential for the peaceful atom was widely promoted in an effort to balance the public image of the atomic program.

Blame for the stall in development of nuclear energy was placed at the feet of Lilienthal and the scientists on the General Advisory Committee (GAC), who were wary of pushing ahead without answering fundamental questions in laboratory experiments. There was also some doubt over the role nuclear power would actually play in providing electricity. In addition to Lilienthal's hopeful but cautious attitude, both Enrico Fermi and Robert Oppenheimer had warned that electricity from nuclear power was "not around the corner." The GAC issued a report that said, "Even on the assumption of a most favorable and rapid technological development . . . we do not see how it would be possible to have any considerable portion of the present power supply of the world enhanced by nuclear fuel before the expiration of twenty years."

This was their most optimistic view, issued after Lilienthal had criticized them for a more dismal assessment. In 1947, the American representative to the United Nations Atomic Energy Committee had said, "I think it will take between thirty and fifty years before atomic energy can in any substantial way supplement the general power resources of the world. That is under the assumption that development is pushed, that intelligent and resourceful people work on the job, and that money is available for it."

The general attitude was that nuclear power would be a "transitional" technology for passing from an age of fossil fuels to one of solar energy. Most estimates of this interim period ranged from 100 to 1,000 years. Some scientists believed, however, that after adequate research, nuclear technology would not prove to be worth the risks it posed and the costs. James Bryant Conant, president of Harvard University and a chemist on the GAC, said that by 1985 he expected solar energy to begin a climb to dominate industrial power production. In a speech in 1951 he did a little crystal ball gazing, predicting that by 1965 "a sober appraisal of the debits and credits of the exploitation of atomic fission led people to decide the game was not worth the candle." He said that although "experimental plants were producing somewhat more power from controlled atomic reactions than was consumed in the operation of the complex process, the disposal of waste products had presented gigantic problems—problems to be lived with for generations." He also cited drawbacks in the great costs of building the nuclear plants.

Other scientists on both sides of the Atlantic echoed Conant's remarks. They reasoned that nuclear power would prove unable to compete with solar energy or other fuels. A 1952 report from the Paley Commission, a group set up by President Truman to study the nation's resources, titled *Resources for Freedom,* predicted that by 1975 America and its allies would be facing the possibility of shortages in fossil fuel. The report urged that solar energy be developed as a replacement for dwindling oil supplies and predicted that, if the nation made such a choice, 13 million solar-heated homes could be built by the 1970s. That would "make an immense contribution to the welfare of the free world."

Interest in solar heating went back to the 1920s and 1930s, and the burst of attention in the early 1950s had evolved from the housing boom following World War II. Experiments were also going on with solar photovoltaics—producing electricity directly from sunlight—a technology that went back to discoveries by the French scientist Edmund Becquerel in 1839. New breakthroughs in photovoltaics in 1954 would herald a new solar age in popular media and *The New York Times* would urge that a portion of the funds going for nuclear development be allocated for solar research.

Unfortunately, recommendations of the *Resources for Freedom* report would be disregarded. Dean and the "developers" on the AEC staff envisioned a nuclear-powered future and were intent on getting an operable reactor in five to ten years. The growing involvement of private industry in the program would help pressure Congress to fund the acceleration of reactor development.

The larger private power companies and chemical companies and major construction companies involved in the atomic energy program had their own agenda for nuclear development. In a speech to the American Association for the Advancement of Science in Cleveland, Ohio, a few days after Christmas 1950, Philip Sporn laid out the industry's interests and where electricity production stood at that point. Sporn was president of the American Gas and Electric holding company and head of the Advisory Committee on Cooperation Between the Electric Power Industry and the Atomic Energy Commission, a follow-up to the first industry atomic study group formed in 1947. He said that one of the major barriers to nuclear development was that "the state of power production and development by conventional means, particularly in this country, is unusually satisfactory." With less than 7 percent of the world's population, the United States was producing 40 percent of the world's electricity. There was also a major building program for conventional plants in midstream. For the period from 1949 through 1952, new electric facilities increased the total supply by close to 290 million kilowatts —more than 50 percent over the capacity installed at the end of 1948. But despite the abundance of electric capacity he described, Sporn told the group it was essential to develop atomic energy in the United States because "it would provide a more universally available, more readily portable and more economical fuel." Then he unveiled a larger agenda.

Individual corporations could not afford the huge amounts of capital it would take to develop nuclear projects. Sporn, head of the largest remaining electric holding company, urged that barriers such as the Public Utility Holding Company (PUHC) Act be revised to let private companies combine resources to develop nuclear facilities.

Gordon Dean was in favor of the idea. By mid-1951 Dean and the other commissioners had approved pilot studies to be undertaken by four industrial groups, combinations of electric and chemical companies. In October Dean told a convention of engineers in French Lick, Indiana, that the new cooperation with industry was "a natural and healthy development, and we intend to do everything we can to work with industry." Dean also alluded to the fact that laws interfering with private participation would have to be changed to allow the pace of reactor development to accelerate. Government assistance to private industry would be substantial. For a study from Dow Chemical Company and Detroit Edison, the federal government would put up $725,000 and the private companies $275,000, which would be passed off in tax write-offs and charges to consumers.

Public power advocates, New Dealers who had fought to control the abuses of the holding companies, and scientists who warned of the uncertainties of rushed atomic development were outraged by the AEC's subsidy and encouragement of the rise of these new private combines. The AF of L issued a report in early 1952 warning of the implications of the shift in policy:

> Atomic energy itself and technology arising out of nuclear development must remain in the public domain. While development of private investment and enterprise for civilian use of atomic energy should be encouraged, such use of it by private enterprise should be strictly competitive. Private monopoly in any phase or segment of the atomic energy industry is intolerable. Yet even now monopolistic aggregates are already building it up, not only around the fringes of this public program but within it. Monopoly of

engineering skill and scientific technology is no less dangerous than monopoly of private capital. No private corporation should be permitted to accumulate materials, equipment, or skill of this industry for its own use or to arrogate to itself the power derived from their exclusive possession.

Others worried about the social and political impacts. In an article in the December 1951 issue of the *Yale Law Journal* one writer commented:

> The major social, economic, and international considerations relating to the development of atomic energy, expressly recognized in the Act, appear to have been forgotten in the . . . urge to instate the enterprise system. And what has happened, one may ask, to the innocent notion that the benefits of atomic energy should accrue to the nation as a whole—without the prior drain of private profits—since the resource itself was brought to fruition by public funds? Somewhere along the circular route of national policy this point got lost. Sooner or later the American people will demand to know where and why.

The public knew nothing at the time, however, and the momentum was clearly on the side of private industry. The democratic process advocated by Lilienthal and Einstein for public discussion and debate had not occurred. But in its place was a process whereby private corporations gained access to top secret information and took part in decisions to shape the future. There was no question of the industry's motives. The Joint Committee on Atomic Energy Report for 1952 said that spokesmen for the private power companies felt an aggressive responsibility to develop atomic power with a dual interest in profits and defense of their future role in the economy. The generating equipment companies, GE and Westinghouse, similarly were said to be interested in "keeping the power consumption rate rising; to the extent that atomic power will abet this rise, they see its development as a natural part of their responsibility to their investors."

Backed by the publicity programs to put a "peaceful" face on the atom and the electric industry's multi-million-dollar propaganda campaign, coloring public ownership in the power industry as "socialistic," the wheels of private atomic development continued to move forward. The constitutional danger of a small government committee controlling nuclear power was now complicated by growing influence of industrial interests—an influence that promised even greater concentration of capital and environmental and social impacts than the holding companies of the 1920s.

The shift that began under Gordon Dean's leadership would accelerate during the Eisenhower Administration. What the private power industry had lost in Wendell Willkie, it found anew in the candidacy of General Dwight D. Eisenhower. During the presidential campaign of 1952, Eisenhower readily picked up the idea of "creeping socialism" in the power industry and took a position against the federal government supplying electric power. Although he pledged in the closing days of his campaign that the TVA would be "operated and maintained at full efficiency," the whole idea of public power was alien to his philosophy and that of his supporters. It was not long after his inauguration that deep cuts were announced in the Valley Authority's budget. And one White House insider, Emmet John Hughes, quoted Eisenhower as saying shortly after his election, "By God, if we could ever do it before we leave here, I'd like to see us sell the whole thing, but I suppose we can't go that far."

Eisenhower's first four years brought a dramatic shock to public power systems. In his efforts to dismantle federal power programs, Eisenhower became embroiled in a controversy over TVA as well known as Watergate would be two decades later. The Eisenhower Administration also rewrote the Atomic Energy Act of 1946, attempted to roll back the PUHC Act of 1935, and essentially handed over the ill-researched nuclear power field to private industry. Atomic Energy Commissioner Lewis Strauss, a conservative Wall Street investment banker and special assistant to the president, was Eisenhower's kingpin in the final wrestling match between private and public interests over nuclear power.

Six months into the Eisenhower presidency, Senator Warren Magnuson told delegates attending the annual convention of the American Public Power Association, "You are in a fight for your very existence." Others viewed as opposed to private industrial interests were also forced to fight for survival. In one of the actions clearing the way for a change in the Atomic Energy Act of 1946, J. Robert Oppenheimer was called before the House Un-American Activities Committee and accused of being a communist sympathizer. Ultimately he was discredited, and his security clearance was taken away. The message to other atomic scientists was obvious: If they could do it to Oppenheimer, they could do it to anyone. There would be no criticism of the Eisenhower Administration proposals from nuclear scientists.

In early 1953, the four study projects undertaken two years earlier were completed. The private consortiums working on the projects all recommended changing the Atomic Energy Act of 1946 to open up the way for private ownership and operation of reactors. There was a very large catch in the way of changing the law, however. Under section 7 (b) of the act, a barrier had been placed against premature private development of reactors. The section read: "Whenever in its opinion any industrial, commercial, or other nonmilitary use of fissionable material or atomic energy has been sufficiently developed to be of practical value, the Commission shall prepare a report to the President stating all the facts with respect to such use, the Commission's estimate of the social, political, economic and international effects of such use and the Commission's recommendations for necessary or desirable supplemental legislation." The president was to transmit the report to Congress with his recommendations. Debate was to follow.

Before he left for a job with the Wall Street firm of Lehman Brothers, AEC Chairman Gordon Dean claimed that such a report was unnecessary because private companies only wanted to study reactors and not develop them for commercial use. True to the warnings of the AF of L and others a year earlier, Dean was backed by large industrial interests. The Dow-Detroit Edison study team had enlarged to include eighteen electric power systems, four manufacturing firms, one chemical company, and three engineering and construction businesses with combined total assets of $8 billion. Added to these were the other three teams, plus GE, Westinghouse, and the large influential bankers associated with these companies. There would be no report until after the act was changed.

Upon Dean's departure in June, Lewis Strauss took over the chairmanship of the AEC and brought to Congress proposals to change the act of 1946 by allowing greater private access to secret technical data, liberal licensing provisions, and private ownership of nuclear reactors. The changes would open the way for private development of nuclear power in the United States and provide the government's blessings to companies such as GE and Westinghouse to sell nuclear technology to other nations.

All through the spring and early summer of 1954, exhaustive hearings were held before the Joint Committee. A fundamental problem was that all the atomic material was still clouded in secret. Without the public report on impacts of the proposed changes, only those in favor of rewriting the law had access to information. Nuclear scientists, of course, had been effectively stifled by the Oppenheimer controversy. The process drew intense opposition.

The CIO criticized the proposal as lacking the candor to be expected from a government body charged with such heavy responsibility as the AEC. Said Benjamin Sigal, its spokesman:

> So far as we know, the Commission has failed to give any consideration to the social, political, and international effects of its proposals, while its discussion of the economic effects is extremely vague and hypothetical. In short the Commission has failed utterly to make a case out of justifying amendments to the act. . . . The CIO is firmly of the opinion that the AEC should never have begun this process of delegating its production responsibility. We believe that the public welfare demands public operation as well as public ownership, of the atomic energy industry. . . .

Clyde Ellis of the National Electric Rural Cooperative Association, which then numbered 1,024 co-ops, criticized the hastiness of the changes. "We do not yet know the full economic, social, and political implications of nuclear power," he said; "our people as yet have had neither sufficient facts nor sufficient time to develop a policy; . . . it is too soon to be changing a law as basic, as vital as this."

Jerry Voorhis of the Cooperative League of America reinforced Ellis' comments. He said, "All that would be needed . . . to pronounce the final death knell over economic freedom would be for Congress to commit the unpardonable breach of public trust of giving to some of the companies which are already in the driver's seat a key to an advantage which might prove unchallengeable. . . ."

Private industry, because of the position it had cultivated on the inside since 1947, was well informed and well represented. The changes were strongly advocated by the Atomic Industrial Forum, an organization of twenty major corporations, including GE, Dow, and Standard Oil, formed in early 1954 to help "advance the development of atomic energy in the best traditions of free competitive enterprise." The president of the organization was Walker Cisler of Detroit Edison. On the board of directors was former AEC Chairman Gordon Dean as a representative of Lehman Brothers.

On August 30, 1954, Eisenhower signed the new Atomic Energy Act, opening the doors for private development of nuclear energy. Combined with atomic bombs, foreign policy analysts saw it as a useful second prong in international relations. In the fall of that year, Representative Sterling

Cole, head of the Joint Committee and a close friend of Strauss's, would write, "Perhaps no single law will affect so deeply our way of life—and that of all mankind—as will the new Atomic Energy Act." Cole had no idea as to how right he would be. For the commitment of government subsidy that would total as much as $200 billion by 1980, the unsolved problems of radioactive waste, and the proliferation of atomic weapons technology there had been no public discussion. It was a fateful decision rushed by a few men, willing to risk public security out of a mixed sense of self-interest, dedication to fulfill the corporate vision of the future, and a naive belief in the ability of technology to resolve fundamental social problems.

The report that had been required before changing the act would not be issued until January 1956. It was nothing less than a bag job. Those who wrote the report were appointed by the Joint Committee, which had recommended the changes in the first place. It was a prime example of the problem of allowing the committee to have both legislative and regulatory powers. In addition to justifying the changes, the report was marked by glaring gaps in assessing the impact on the economy and the structure of the electric power industry. No mention was made of the impact on the financial community (which would make billions of dollars from the capital to be raised building nuclear plants). No mention was made of the growing concentration of the power industry. No mention was made of the historic battling among the public, cooperative, and private sectors of the industry. A single sentence broadsided any meaningful examination: "Despite the importance of this new measure," the authors wrote, "it appears that the use of atomic power will cause no major changes in the character or pattern of the industry." The report was not the only fiasco. This back-room political dealing was overshadowed by the revelation of a scandal involving key officials of the Eisenhower Administration.

At the same time the Atomic Energy Act was moving toward passage, Eisenhower's efforts to dismantle federal power programs burst onto the front pages. In June 1954, Eisenhower ordered the AEC to purchase 600,000 kilowatts from two private holding companies, the Southern Company and Middle South Utilities (MSU)—the remains of the Commonwealth and Southern holding company after the breakup forced by the New Deal PUHC Act. Eisenhower was in favor of the Dixon-Yates deal, named after the presidents of the two companies, because it would eliminate the need for TVA to build more plants to serve the AEC facilities at Oak Ridge. He wanted the new plants to be built by private power companies. The issue touched off a storm of protest and an embarrassing scandal.

Senator Estes Kefauver, a Tennessee Democrat, saw the proposal as the "entering wedge," which the private utility lobby had been striving for in its campaign to eliminate public and cooperative competition from the electric power business. In an investigation by the Senate Judiciary Committee, Gordon Clapp, the former TVA director, charged that "the Dixon-Yates deal would lead to the formation of a new combine of two large holding companies, each with a long history of aggressive hostility to TVA, acting in concert with two government agencies, the Bureau of the Budget and the AEC, to squeeze the TVA into submission to the private utilities."

The most revealing testimony heard by the committee came from James D. Stietenroth, the former secretary-treasurer of the Mississippi Power and Light Company. He was fired from his job within hours after publication of a sensational public statement in which he supported what Clapp had said and charged "Wall Street domination" of the issue. Stietenroth told the committee that the proposed Dixon-Yates combination would enlarge the power and influence of the two holding companies and boost their combined assets to $2 billion. "It is so huge," he said, "it is frightening. . . . I am actually frightened as a citizen of the United States by the bigness that even my company and my group of companies represents."

MSU subsidiaries, he told the committee, were still dominated by an absentee holding company, EBASCO Service Company (the remains of the Electric Bond and Share holding company controlled by Morgan interests), with offices at 1 Rector Street in New York City. Local company officials, including boards of directors, he said, were little more than puppets who signed papers drawn up at the New York holding company headquarters. In response to the question as to how this situation differed from the condition before passage of the PUHC Act, Stietenroth said, "Sir, except that we have by and large legalized it and gotten Securities and Exchange Commission approval pretty well down through the line, I don't see any difference. It is the same thing just wrapped up in a different package."

The revelation rocked Congress. The power industry, assumed to have been restructured by the PUHC Act, was still acting in concert with Wall Street firms and exerting substantial influence over the federal government. Eisenhower found himself embroiled in the controversy. A succession of testimony given before the committee revealed that the White House had pressured the Federal Power Commission to approve the

Dixon-Yates contract and that Sherman Adams, the president's assistant, had ordered a recess in SEC hearings at a crucial time—the very day that Adolphe H. Wentzell was to have appeared as a witness. Wentzell was a prestigious financier who proved to be the key to the inner machinations. Along with being a part-time consultant to the White House Bureau of the Budget, he served as a vice-president of the First Boston Corporation, a financial agent for Dixon-Yates. In one part of his dual role he had prepared a secret report for the White House on how to curb TVA's power policies. And as an executive for First Boston, he played a part in the formulation of the Dixon-Yates contract, a role which was at first publicly denied but finally admitted.

On July 11, 1955, Eisenhower was forced to order cancellation of the contract. Public power proponents were jubilant that they had been able to block the attempt of private power companies to cross the Mississippi River into territory they had been evicted from twenty years before. The Dixon-Yates controversy was only one of the campaigns Eisenhower was waging on behalf of private power companies.

In August 1955, a huge blow was struck against rural cooperatives and public power systems in Idaho. The FPC finalized a decade-long battle by granting approval to Idaho Power Company to three dam sites on the Snake River, including a location known as Hells Canyon. A 792-foot dam for Hells Canyon proposed by the Bureau of Reclamation would have been "the most impressive dam in America" according to Clyde Ellis of the Rural Electric Cooperative Association.

In other regions of the country public power systems and cooperatives were being isolated and squeezed out by the Eisenhower Administration's rules on federal power transmission. In 1956 a House investigating committee would report: "Commencing in 1953 the new officials in the executive departments began a determined effort to sabotage and wreck the Federal power program." According to the investigation, changes in federal power rules had been taken directly from a document prepared for Eisenhower officials by a lobbyist of Pacific Gas and Electric Company. The report also led to an investigation of private power companies' efforts to influence the Secretary of the Interior.

At the same time these fights were taking place within the walls of government, public power companies were witnessing a propaganda and advertising campaign similar to what had gone on in the 1920s. Against the background of the McCarthy anticommunist hysteria, private power companies helped fund and distribute material from the John Birch Society in an effort to undermine consumer support for public power. Just as corporate spokesmen had charged as far back as the 1890s, pamphlets, speakers, and films stressed that public power was the first step to socialism. Alex Radin, general manager of the American Public Power Association, told delegates at the twenty-third conference of California Utilities, "I am convinced an investigation would show the propaganda, lobbying and other such activity of the private power utilities exceed those of the Insull-Hopson era."

Even Eisenhower took part in the propaganda efforts of the private power industry.

□ □ □

A week after he had signed the Atomic Energy Act of 1954, President Eisenhower appeared in an unusual national television broadcast from a Denver, Colorado, studio. In his hand he held a 2-foot-long "wand" containing a neutron power source. Half the continent away outside Pittsburgh, Pennsylvania, officials stood at the site of the Shippingport nuclear plant, anxiously watching silent robot-controlled construction equipment. With millions of viewers tuned in, Eisenhower waved the wand over a fission detector. A needle on the detector swung, an electric current raced through the wires, and a steam-shovel at the site started automatically and roared out before the officials and press, literally breaking ground for the plant. Strauss later laughed at the event dreamed up by power company public relations men.

Shippingport was to be the nation's first commercial reactor, and the gimmick was meant to symbolize the new era of the atomic wizardry and the beginning of the atom's "service" to mankind. In a more telling way, it symbolized the cheap magic the industry would use to continue dazzling the public as questions of safety and cost of the atomic technology were swept aside.

Strauss had negotiated the deal for the Shippingport reactor as a joint venture of the AEC and Duquesne Light Company, even before changes in the act had been passed. He saw Shippingport as a crucial first step in the "civilian" reactor program. His enthusiasm was evident in the fact that a definitive contract was not written until after the ground breaking. It would establish Westinghouse as the principal contractor and place Admiral Hyman Rickover, who had overseen the development of a reactor for the atomic-powered submarine *Nautilus* the year before, in charge of the project. Although Strauss praised Duquesne Light in glowing patriotic terms as the first electric company to enter the nuclear age, the federal government would pay as much as 75 percent of the costs for the project.

In January 1955 the bureaucratic path made by the Shippingport project was broadened in a five-year plan known as the Power Reactor Development Program. For this program the AEC solicited proposals for prototype plants that previous studies had selected as the best bets for economically feasible power reactors. It offered a tremendous bargain for private power companies. The federal government would pay research and development costs, and the companies could pass their remaining costs to consumers. Responses came for development of four of the five prototypes. In the Northeast, Yankee Atomic Power Company proposed building a 134,000-kilowatt reactor at Rowe, Massachusetts. A fast breeder reactor would be built at the government experimental station in Arco, Idaho. A sodium reactor experiment was planned for Santa Susana, California, by North American Aviation, Inc. And a reactor combining and testing various technologies was planned for the commission's Oak Ridge operations.

Requests for a second round of proposals for smaller reactors ranging in size from 5,000 to 40,000 kilowatts brought plans from several rural cooperatives, universities, and city power departments including Holyoke, Massachusetts, and Orlando, Florida. Out of this second round would come small experimental plants for the Rural Power Cooperative Association in Elk River, Minnesota, the Consumers Public Power District in Hallam, Nebraska, the city of Piqua, Ohio, and the Philadelphia Electric Company's Peach Bottom plant.

Subsequent proposals for large reactors came from Consolidated Edison of New York for a plant at Indian Point, twenty-four miles up the Hudson River from New York City, Commonwealth Edison for the Dresden plant at Morris, Illinois, General Electric and Pacific Gas and Electric at Humboldt Bay, California; and in the wings was a proposal from the Power Reactor Development Corporation (PRDC) led by Detroit Edison for a breeder reactor to be built on the west shore of Lake Erie at Monroe, Michigan.

These would be the nuclear plants of the early 1960s. The visions of these plants as benign servants of mankind was illustrated in an article on Consolidated Edison's Indian Point proposal. Sprawled across two pages of the August 1955 issue of *Popular Science* is the sketch of a magnificent complex with a reactor building set deep in bedrock beside the Hudson River. To calm public fears over the plant's safety, the article explained that the pressurized water reactor was one that engineers regarded as "inherently safe." In tests, the reactors were described as being forced to run away "only with great difficulty" and in those cases the "fuel elements simply melt and the fire goes out." Even if the "impossible catastrophe" of a serious malfunction struck, readers were assured that it "would be serious but not excessively dangerous. For the furnace will be underground, in a hole blasted out of solid rock." To dismiss any final doubts the article explained: "No one expects any of these things could possibly happen. They are about as likely as collapse of the Empire State Building. That they have been planned for is a measure of the extreme care Government and industry have given to the establishment of the atom as our servant of tomorrow."

While the public was being told that accidents were unlikely, private companies were demanding some kind of indemnity program to ensure they would not be held responsible for possible widespread damage from nuclear accidents. This fear—combined with questions over the potential for power companies to form conglomerates and escape the oversight of the PUHC Act of 1935—raised huge barriers in moving ahead with construction on the reactors that had been approved.

Strauss and the private power supporters on the Joint Committee sought to eliminate these barriers with two landmark pieces of legislation: one to break down the PUHC Act and the other to provide government-subsidized insurance to protect the industry from claims in the event of a nuclear accident. In the meantime, congressmen who supported public power charged that the atomic construction program was stalled indefinitely and the United States would fall behind in what was now seen as a "nuclear power race" with Russia. Representatives Albert Gore of Tennessee and Chet Holifield of California introduced a bill authorizing a crash program for the AEC to build federal power reactors. Through all of 1956, legislative proposals for federal reactors, nuclear insurance, and gutting of the PUHC Act of 1935 set off far-reaching discussion in the back rooms of Congress.

In April and May 1956 several hearings were held on the proposed Electric Energy Development Act of 1955, which attempted to exclude from the definition of "holding company" any organization involved directly or indirectly in the "ownership or operation in whole or in part" of nuclear power facilities. Introduced by Senator Charles Potter of Michigan and Senator John Pastore of Rhode Island, the bill reasoned that the nation's general welfare and national security were promoted by an expanding economy, which "must in turn be promoted, among other means, by encouraging the development of an ever-increasing supply of low-cost electric energy. . . ." Loosely translated, this boiled down to "What's good for General Electric and the power companies is good for the nation." In practical terms, it meant that brokerage firms, large banks, GE, Westinghouse, or large oil corporations would be able to finance the projects and corner the market on atomic energy, and that the holding company systems could rise up again in the form of huge nuclear "generating companies."

Walker Cisler, the president of Detroit Edison and the head of the Atomic Industrial Forum, told Pastore's hearing panel that the consortium he had formed with Dow Chemical and twenty-five other companies, the PRDC, would not go ahead with its plans for a breeder reactor in Senator Potter's home state of Michigan unless the PUHC Act was amended. Senator Pastore expressed extreme concern, making the whole fate of nuclear development dependent on Cisler's success. Pastore offered the bill's critics a vision of the future concocted in the board room of the New England Electric System holding company, which operated in his home state. He naively said, "If this thing clicks, if we can get to the point that through the use of atomic energy we can produce electric power competitively with the use of conventional fuels, you won't have to worry about preferences anymore. We will have it coming out of our ears. That is the hope of the future. . . . If we can produce power by atomic energy, maybe we won't even need meters. It will be so cheap to produce that the cost would be only in transmittal."

Despite Pastore's wrangling, Cisler's threats, and pressure from other power companies, the bill to gut the PUHC Act was defeated.

At the local level, Cisler's breeder reactor was also facing difficulties. In June 1956, just a month after the PUHC Act hearings, the Advisory Committee on Reactor Safeguards—a group of scientists and engineers working on the siting of new reactors—issued a surprising report to the AEC. In regard to Cisler's proposal to build the Fermi Breeder Reactor outside Detroit, the committee said, "There is insufficient information available at this time to give assurances that the Power Reactor Development Corporation reactor can be operated at this site without public hazard." Strauss didn't know what to do with the report. A defeat for Cisler, head of the Atomic Industrial Forum and the nation's largest consortium planning to build a reactor, would have posed a severe blow to the private reactor program. He solved the problem by classifying the report as "secret" and filing it. Outraged when he discovered this action, Commissioner Thomas Murray went before a House subcommittee and revealed what Strauss had done.

Strauss was severely criticized by both Representative Chet Holifield of California and Senator Clinton Anderson of New Mexico, who cochaired the Joint Committee. A showdown between Strauss and Anderson demonstrated, however, that the AEC would regard the Joint Committee only as another advisory group on issues of licensing reactors.

Strauss explained that the final decisions on siting and safety lay with the commission.

On August 4, 1956, a permit was granted for Detroit Edison's Fermi Breeder Reactor, a 150,000-kilowatt plant planned to cost $40 million. Four days after the permit was granted, Strauss arrived to give a speech at the preplanned ground-breaking ceremonies. It was apparent that he had greased the skids for the plant. But in the momentum Strauss was trying to build and beneath his roughshod style, deeper trouble was brewing.

While many people swallowed the *Popular Science* and common Sunday magazine pabulum on the rosy future promised by nuclear power, a growing number had serious misgivings about nuclear power, particularly if they were faced with living near a reactor. As early as 1948, Congressman Melvin Price of Illinois said he had to assure distraught landowners in the area around the AEC reactors at Argonne, Illinois, that they were not endangered by the plants. And in the West, particularly in Nevada, where the commission's expanded weapons program had begun testing atomic bombs in 1951, stories had circulated of towns being plastered with radioactive ash and people being cautioned to stay inside their homes for hours at a time because the radiation levels were too high. By the mid-fifties, stories had also begun to include ranchers' tales of high numbers of sheep being born deformed or dead.

In Detroit and the area surrounding the Fermi site at Monroe, Michigan, people questioned not only atomic technology, but the possibility for human error in building and operating the reactors. Many were upset enough by the project to consider taking the PRDC to court in an attempt to block construction. In Washington, Senator Anderson, stung by Strauss's arrogance and fearing the political fallout of an accident near Detroit, contacted Michigan Governor Mennen Williams and UAW President Walter Reuther in an attempt to fan the flames of opposition. Although Williams would step away from the issue, Reuther and the UAW would take it head on.

Strauss would later remark that, although there had been public concern about construction of the Shippingport plant, it wasn't as serious as the opposition to Fermi. The Fermi battle, beginning in 1956, Strauss said, was "the first indication that the private development of atomic power would be fought." Leo Goodman, an MIT-trained engineer and an organizer for the UAW, had spent time at Oak Ridge in the late 1940s and was familiar with radiation hazards faced by workers. He became the union's lead man on the issue and over the decades would become the "grandfather to the antinuclear movement."

The UAW first tried to persuade the AEC to withdraw the plant's construction permit. Denied that request, they took their case to court. The suit was centered on a fundamental gap in the nation's nuclear laws. The licensing of a reactor was done in two parts, first a license for construction, then a license for operation. The AEC argued that they could grant a construction permit without first deciding whether or not a plant would be safe. According to Strauss, the safety issue would be decided once the plant was complete and ready to license. UAW lawyers claimed that this rationale was insane. There would be little likelihood of the commission refusing an operating license once a multi-million-dollar nuclear plant was completed. The case would work through the court system for five years. Although the UAW won lower court rulings, the Supreme Court, in a 7–2 decision in June 1962, favored the AEC. In October 1966 the Fermi plant would suffer a partial meltdown, seriously endangering the Detroit area, but in the dreamy days of 1956, that event was a decade ahead.

□ □ □

The enthusiasm shown by Strauss for nuclear development was shared by his colleagues on Wall Street. The progress of the civilian reactor program turned up great interest within the financial community. At the third annual meeting of the Atomic Industrial Forum (AIF) in September 1956, Paul Genachte, director of the Atomic Energy Division of the Chase Manhattan Bank, said that atomic energy would come to "encompass all the industrial activities of the nation. It will, in time, cut across our entire industrial and economic pattern. In time, also, each existing industry will become concerned with some phase or some particular aspect of atomic energy." Chase Manhattan, like other major institutions, was eager to get in on the ground floor of financing the new technology. Genachte cited a study showing that as much as $7.5 billion in financing would be needed between 1955 and 1965 for nuclear power's research and development. Most significant was the fact that construction of nuclear plants would be more expensive than construction of coal- or oil-fired plants. This assured an expanded role and increased profits for Wall Street. Genachte assured the AIF audience that there would be no problem raising the money to finance nuclear plants.

Support for Genachte's claims came from Newton I. Steers, president of the Atomic Development Mutual Fund. A few months earlier, Steers had described to a congressional panel the phenomenal growth his fund had experienced. In the first year of operation, its capital had risen from $1.25 million to over $15 million. Steers not only marveled at his personal success but at how this reflected "the public's view of the financial future of atomic energy. . . . The investing public evidently agrees with us that atomic stocks have a greater growth potential than any other group of stocks which could be assembled."

The influence of these investors and financial institutions, long tied to the heavy financing of the electric power industry, were clearly having an effect on the course of nuclear reactor development. In 1954 they provided significant leverage in changing the Atomic Energy Act through Gordon Dean, who went from the AEC to the brokerage firm of Lehman Brothers, and through Strauss as his successor. In late 1956 the financial community, a powerful coalition of electric power companies, Strauss, and Eisenhower worked to defeat the Gore-Holifield bill, which proposed a "crash program" for the government to build reactors at six sites. Passed by the Senate but defeated in the House, the bill was said to have brought back memories of vicious New Deal legislative battles over the power industry two decades before. Defeat of this bill put private power companies in the driver's seat for development of nuclear power.

Congressmen on the Joint Committee pushed Strauss for faster progress in the civilian reactor program. In response, Strauss called for immediate private construction of two large-scale reactors and three smaller reactors. Although the economics of nuclear power were largely unknown, Strauss played on the private power companies' fears by threatening to request authorization for government-built reactors if private industry did not build the new plants to be operational by June 30, 1962. Five groups responded: the Florida Nuclear Power Group (Tampa Electric Company, Florida Power and Light, and Florida Power Corporation); Pennsylvania Power and Light and Westinghouse; Carolina-Virginia Nuclear Power Associates (Virginia Power Company and Duke Power); an Ohio group headed by American Gas and Electric; and Pacific Gas and Electric.

In order to make the proposed plants viable, Strauss and members of the Joint Committee pressed ahead with plans to provide the insurance the power companies demanded before they would go operational.

In September 1956, Representative Melvin Price laid out the plan for a massive federal subsidy to insure private operation of nuclear power plants at an Atomic Industrial Forum gathering. During the previous six months Price and other members of the Joint Committee had been meeting with representatives of insurance companies and power companies to write legislation for the necessary coverage. It amounted to the largest governmental subsidy ever given to private industry. The insurance companies would form syndicates to insure the plants. Coverage would be limited to $560 million, of which the federal government would cover $460 million. Any claims above that amount would go unmet.

Price posed a very candid question to his audience: "Are these measures genuine aids to a fledgling private enterprise? Or are they the means of entrenching monopolies which, in the words of Commissioner Murray, desire to 'stake out' atomic energy as their private domain? I haven't made up my mind on these questions, and I don't believe a majority of the members of the Joint Committee or Congress have." They were good questions, but Price provided only a common flag-waving answer. Despite any misgivings he harbored, Price told the group in his next breath that he was committed to private reactor development because "our national interest and international prestige require we support a policy of accelerated reactor development. . . ."

The Price-Anderson Act, the government's nuclear insurance legislation, was offered to Congress in the spring of 1957. At the same time a report was issued by the Brookhaven Institute on the possibility and potential impacts of a nuclear accident. Contracted for by the AEC, the report, *Theoretical Possibilities and Consequences of Major Accidents in Large Nuclear Plants* (later known simply as *WASH-740*), was supposed to provide financial and geographic dimensions of an accident at a nuclear plant and facilitate the passage of Price-Anderson. However, it ended up stunning the nation with the picture of devastation it portrayed. Although the likelihood of an accident was given spongy astronomical odds ranging from one chance in 100,000 to one in a billion, the report showed that a total area of some 150,000 square miles could be affected by an atomic accident. On the basis of a "runaway" reactor at a theoretical plant located thirty miles from a major city, the report estimated that 3,400 people living within fifteen miles of the plant could be killed and another 43,000 injured. Property damage could range as high as $7 billion. The fears this image of devastation conjured up helped push the massive insurance subsidy through Congress. It also provided nuclear opponents with data they never had before and spread public resistance to construction of nuclear plants.

In October 1957, public fears were reinforced further by news of a nuclear accident at the Windscale nuclear facility in Great Britain. A *Business Week* reporter wrote, "Every new detail about Britain's Oct. 10 runaway of an atomic reactor seems to emphasize that the nation escaped a major catastrophe more by good luck than good management. And the British public is getting increasingly indignant about it." At the onset of the incident, with a fire raging out of control inside the plant, a neighboring reactor was shut down and 3,000 workers evacuated. Police were notified to prepare to evacuate the area's population, but public disclosure was withheld to prevent panic. After three days the fire was finally brought under control, but because of the release of radiation from the plant, milk from some 1,000 farms in a two-hundred-square-mile region was later confiscated and dumped. Britain's Atomic Energy Authority, which projected two thirds of their electricity would be generated by nuclear power by 1975, was accused of playing down the seriousness of the accident.

There had been other serious accidents, too. In Canada the Chalk River reactor had undergone a partial meltdown in December 1952 and released a million gallons of radioactive water. Several small accidents in the American program had caused radiation-related deaths of seven employees. But in 1957 the AEC continued to claim that in thirteen years of work, "no member of the public" was known to have suffered overexposure to radiation.

On the other side of the issue, the drive for rapid commercialization of nuclear power was given greater impetus by news that the Soviet Union had launched Sputnik—the first satellite to orbit the earth. The space race now joined the nuclear race, and Strauss and private industry continued to press ahead, despite doubts over safety or economics.

In December 1957, after a rapid three years of construction, the Shippingport reactor came on line as the nation's first commercial nuclear plant. In the construction pipeline were several other commercial reactors. On leaving the AEC in 1958, Lewis Strauss proudly recalled the great expansion that had taken place. "In 1953 there were in existence only two small power reactor experiments," he said. "Five years later, eight civilian power reactors and large reactor experiments were in operation or under construction." Strauss's policies and legacy at the AEC would leave a long trail. Of the forty-three reactors in operation in 1974 only four would be owned and operated by public power or cooperative systems. Private power companies, with their massive influence, had secured what would turn out to be a very volatile corner on electricity from nuclear reactors. As those first plants came on line in the early 1960s, the public would demand a more objective assessment of the questions surrounding nuclear power. Harsh new economic and environmental realities were about to dim the dazzle of the technology and promises of unlimited abundance.

5. *Limits to Growth*

The age of innocent faith in science and technology may be over. We were given a spectacular signal of this change on a night in November 1965. On that night all electric power in an 80,000 square mile area of the northeastern United States and Canada failed. The breakdown was a total surprise. For hours engineers and power officials were unable to turn the lights on again; for days no one could explain why they went out; even now no one can promise that it won't happen again.

Barry Commoner
Science and Survival, 1966

The 1960s and early 1970s brought about a series of sudden confrontations and shocking realizations for Americans. Out of the shadows of the Cold War a conflict in Southeast Asia rose to haunt a generation and transform the global image of the United States. Inflation began a seemingly endless climb. The political, if not physical, limits to resources were dramatized by the Arab oil embargo. And in every locale the growth of environmental awareness and the recognition of deep social costs of technology shook industrial giants to their foundations. No industry was hit harder by this than electric utilities and the dream of endless abundance they had been selling for decades.

The 1960s began full of promise. The election of John F. Kennedy and the launching of his New Frontier program proclaimed a stretching of all the old boundaries of prosperity. The United States would mount unparalleled efforts in the space race with Russia, and in the nuclear race, to maintain global technological dominance. Electricity was viewed as a key to these efforts and other industrial expansion. In Kennedy's vision, ancient regional conflicts and struggles between public and private interests would be laid aside in order to utilize fully the nation's resources.

A month after taking office, Kennedy urged accelerated production of energy from nuclear sources. Citing Federal Power Commission estimates, he said total installed electric capacity should triple by 1980 if the nation was to maintain essential economic growth. "Sustained heavy expansion by all power suppliers—public, cooperative, and private—is clearly needed. . . . We must begin now also to plan for regional cooperative pooling of electrical power. Both efficiency and growth goals will be served if we interconnect our hydroelectric and thermal power resource plants."

There was a darker side to this hope. Within a few years, Kennedy's plans would bring about a new era of centralization in the power industry and literally change the face of public power—drawing it into the fold of a re-emerging power empire. It would set the stage for accelerated reactor development and widespread protests. The first emergence of these events took shape in the "power pools."

Power pooling—linking up the transmission lines of power companies in regions—was an idea that went back to the Superpower and Giant Power plans of the 1920s. Over the decades, battles for control of service territory had slowed the growth of interconnections and their benefit to consumers. But plans for huge nuclear plants and the formation of regional conglomerates of utilities to build them sparked the growth of power pools and a rash of new interconnections.

At the time, the nuclear program was perceived to be moving sluggishly at best. A *Newsweek* magazine reporter wrote, "The big reactor program, which had captured the public imagination with the millennial visions of smokeless, all-electric cities, suffered most from rising costs and technical drag-outs. The Dresden, Illinois, nuclear plant 50 miles from Chicago and the Yankee Atomic Electric Plant at Rowe, Massachusetts, are generating electricity. But the Enrico Fermi plant at Monroe, Michigan, has been delayed several times." While Kennedy pushed his appointee to the Atomic Energy Commission, Glenn Seaborg, to accelerate the reactor program, new questions arose over who would control the power pools.

As in the old days, power pooling allowed large private power companies in some areas to squeeze out smaller competitors, including cooperatives and municipals. In other areas it brought about a truce between public and private power companies to exploit power resources jointly. In early 1961, for example, two private power companies, Northern States Power and Otter Tail Power Company, and three rural cooperatives revealed plans to form a power pool in North Dakota and build five 200,000-kilowatt plants by 1977. When the plans were announced in

1961, only four power pools existed. By 1970 there would be seventeen power pools, representing half of the nation's electric capacity.

Many local public power system managers were reluctant to take part in these joint ventures. But amid warnings that they would be swamped in a sea of privately controlled nuclear power, they were urged to "tie or die." In May 1961 at the annual convention of the American Public Power Association in San Antonio, Texas, power pooling and a "new era" of power supply from nuclear plants were the main topics of conversation. After eight years of Eisenhower's actions to dismantle federal power programs, public power systems were seriously debilitated. D. J. DeBoer, the association's president, told the 600 members gathered at the convention that power pooling "may be the only means for continuation of local independent operation." They were urged to trust federal leadership and not give up hope for publicly controlled nuclear power.

Representative Chet Holifield, who had cosponsored the controversial legislation to build federal reactors five years before, told the group he still did not believe that the nation's atomic program should be primarily dependent on private nuclear projects. "I suggest that you prepare for the future use of atomic power by planning now for interconnection of your public and co-op systems," he said.

Local public power managers remained wary, however. Connecting into a power pool was a major step away from local control by their consumers. But pressed by competition from private systems and issues of survival, they joined larger public consortiums and tied into the pools controlled by private utilities. As a result fundamental concepts of local control and service to consumers would take a back seat to the initiatives set by private companies. They were "captured." In a few short years public power officials would find themselves aligned with private power executives in debates against angry environmentalists. Later they would take positions against their own consumers over the construction of massive power lines or investment in nuclear plants.

More than a few individuals held out hope that the power pools would become part of a federal transmission system. If such a development came about, it would tie the nation's electric plants together and create a balance of control over the industry, replacing the "yardstick" concept that had been destroyed during the Eisenhower years. At a meeting on the ratification of the United States–Canada treaty for upstream development of the Columbia River, Alex Radin of the American Public Power Association said he looked forward to a time when we "think in terms of the interrelationship of power resources of Alaska, Canada, the Pacific Northwest and the Pacific Southwest. . . ." At the same event, UAW

Vice-President Pat Greathouse called for a national power grid developed and maintained by the federal government. Clyde Ellis of the National Electric Rural Cooperative Association proposed a national grid jointly owned by public and private interests. Already bits and pieces of such a grid were forming in the pools. The Colorado Basin Consumer Power Group, including public systems from Colorado, New Mexico, Arizona, Utah, and Wyoming, was pressing for a regional federal transmission system. And in the Northwest, the Bonneville Power Administration was extending its lines into California to form the backbone of a West Coast grid. Private power interests met the proposals for a federal or a jointly owned national transmission system with fierce opposition.

Still locked in their Cold War "anticommunist" strategies to eliminate public power systems, private power companies took advantage of Kennedy's policies to combine their interests and position themselves to take greater control of geographic regions. They wanted control of transmission, and they wanted control of nuclear plant ownership. Edward Vennard, head of the Edison Electric Institute, campaigned against what he called the socialistic dangers of a federal transmission system. He also strongly criticized Kennedy's support for a new proposal for federally owned nuclear plants.

In March 1961, President Kennedy set the private power companies reeling by requesting $60 million for development of a generating plant to use the waste heat from a plutonium reactor in the weapons program at Hanford, Washington. As the first steam plant to be run by the government outside the Tennessee Valley area and the first government reactor to supply electricity, Vennard saw it as "the first link in a coast-to-coast grid tying in all of the regional power systems." Echoing the ancient propaganda of the war between public and private interests, he called the bill "contrary to the public interest" and against "American principles." For the next year, the proposal set off a running controversy over whether it was an efficient use of waste heat or the "first step to socialism."

Previous presidents of Edison Electric Institute had made accusations in more biting terms. At the 1959 EEI convention President J. E. Corette of Montana Power Company said, "Government ownership of utilities has always been the first goal of the socialists and communists. Because of this, the future of the American system of government is dependent on the electric business continuing in the hands of investor-owned, taxpaying companies. . . . Our problem is not only to save our industry, but to save the American system of government."

While the bill for public power systems to use the reactor at Hanford would fail under heavy lobbying in 1961, the following year his supporters and public power advocates drove it through. The catch was that the federal government would not own or operate the plant. The Hanford generating plant was designated to a small group of public power systems known as the Washington Public Power Supply System, which had joined together to build a dam in 1957. Far from bringing socialism to the Northwest, two decades later the seeds of ambition sown here would be manipulated by private industry and would bring about the creation of a massive five reactor construction project that would collapse in the nation's largest municipal default.

□ □ □

During the time that private power companies were working against the new proposal for federal reactors, other events occurred that shook the electric industry to its core. In Senate hearings during April and May 1961, extensive violations of the Sherman Antitrust Act were revealed to have been committed by executives of GE, Westinghouse, and several other electrical manufacturers. Following a grand jury investigation, a stream of executives appeared before a Senate committee chaired by Estes Kefauver. They confessed to clandestine bid-rigging operations in which equipment prices were inflated, bids were agreed upon in advance, and contracts were divided up according to set percentages. Westinghouse participation in the deals amounted to over $1 billion. It was significant that the TVA had blown the whistle on the practice. Private power companies had passed the inflated costs for equipment on to their customers with no complaint. Some of the bid rigging had been going on for twenty-five years, but the bulk of it occurred during the Eisenhower years.

Ultimately forty-two executives and thirty-two companies were indicated—an act taken in some corners as an indictment of the free enterprise system. Several top-level executives received fines or brief jail sentences. One committed suicide. President Kennedy called the affair a "shadow on the shoulder" of American business.

Kennedy fought to restore public power as a competitive force in the industry and remained opposed to a monopoly of electricity by government or by private companies. Like Roosevelt he saw an abundance of electricity as bringing about a new level of civilization. In August 1962 he appeared at the dedication of the Oahe Dam in Pierre, South Dakota. From a platform on the bank of the Missouri River, he marveled at the dam and talked of the "miracles of engineering," wondering at the impact of electricity on modern life.

> When we are inclined to take all of these wonders for granted, let us remember that only a generation or two ago all the great rivers of America: the Missouri, the Columbia, the Mississippi, the Tennessee, ran to the sea unharnessed and unchecked. Their power potential was wasted. Their economic benefits were sparse. And their flooding caused an appalling destruction of life and property. Then the vision of Theodore Roosevelt was fulfilled by Franklin Roosevelt. . . . And as a result this Nation began to develop its rivers systematically, to conserve its soil and its water, and to channel the destructive force of these great rivers into light and peace. . . . Less than thirty years ago, in the lifetime of most of us here, as you know, fewer than 10 percent of all our rural homes in this country had electric power. . . . That's how quickly the face of a nation can be changed by determination and by cooperative action by all the people. . . . Today over 95 percent of rural homes have electric power.

While the public remained hopeful of the bright prospects for the future, a harsh reality was also creeping in from the edges of Kennedy's New Frontier. In areas where nuclear reactors were proposed, pockets of dissent began to emerge. In the summer of 1962, opposition erupted over Pacific Gas and Electric's proposal for a nuclear plant on Bodega Bay in northern California. And on the East Coast, Consolidated Edison of New York kicked over a political hornets' nest by proposing to place a nuclear plant in the Ravenswood district of Queens, a mile from the heart of New York City.

Con Ed's proposal came on December 10, just twenty days after the release of a controversial AEC report. President Kennedy had requested that the commission take a hard look at the timing and scope of the nation's nuclear program. Aside from magnifying the glittering hopes for nuclear power, the report noted "for safety reasons, prudence now dictates placing large reactors fairly away from population centers." Con Ed's proposal for Ravenswood was a challenge to this AEC safety recommendation. Voices rose up on all sides.

David Lilienthal, gone a decade from the AEC, told reporters "I would not dream of living in Queens if a huge nuclear plant was located there." He also called for an end to government subsidies for private nuclear development and criticized the commission for trying both to regulate and to promote nuclear power. Lilienthal was attacked by the Atomic Industrial Forum (AIF) and others who supported the Ravenswood nuclear plant.

The controversy over the reactor was debated in the New York City Council for a year. In June 1963, an eight-hour-long city council hearing took place over an ordinance to bar construction of the plant. More than fifty witnesses spoke. Many of the opponents referred to the 1957 Brookhaven report, assessing the probability of accidents. It was a raucous meeting. A group of women opposing the plant sat in the balcony wearing yellow paper badges imprinted with a fallout shelter symbol and bearing the slogan "Danger: Radiation—Ravenswood Nuclear Plant." The company's testimony was all but drowned out in the chatter and noise, and occasionally speakers were interrupted by derisive shouts.

Leading the official opposition, state Senator Seymour Thaler of Queens told the council that four out of every five of the borough's residents vehemently opposed construction of the nuclear plant. He reasoned that despite any economic advantages it was unthinkable to endanger the lives of 10 million people. "The mind of man," he said, "has not yet invented an accident-proof piece of mechanical equipment. If the Tacoma Bridge could fall down, if the *Thresher* nuclear submarine could sink, if the Mercury space capsules can be faulty, and if a large part of Manhattan Island could have a blackout in June of 1961, then certainly nobody can be absolutely sure about the safety of the Ravenswood project." Supporters of the plant emphasized economic advantages of siting it in the city and stressed safety features. The city's trade union and construction leaders spoke in favor of the proposal, saying that the city could not afford to turn down a $175 million project.

A letter from AEC Chairman Glenn Seaborg shocked city officials with the news that they had no authority to stop construction of the plant. Seaborg wrote, "The act and its legislative history make it clear that Congress intended that the licensing and regulation of nuclear reactors, for purposes of control of radiation hazards, is to remain the exclusive responsibility of the Federal Government." Contrary to their previous warnings about the danger of federal control and "socialism," private power companies supported this kind of undermining of the democratic process.

Faced with such strong local opposition, however, Con Ed was eventually forced to withdraw its plan in late 1963. It was a significant defeat for the electric company Thomas Edison had given birth to and for the entire industry.

Elsewhere, opposition to other plants appeared. In addition to the criticism of the Bodega Bay project, Californians also opposed a 325-megawatt plant proposed for San Clemente by Southern California Edison and a 490-megawatt plant proposed for Malibu by the Los Angeles public power department. The California Democratic Council, the liberal wing of the state's Democratic Party, opposed construction of the plants out of concern over the danger of accidental release of radioactive gases and the existence of geologic faults near the sites. The early resistance to nuclear plant construction was spreading. The AIF noted that, while local and state officials had gotten involved in the fights, congressmen were steering clear of the issue. But at the local level, officials of the companies observed that opposition was beginning to show signs of becoming a national grass-roots movement. A spontaneous new effort at public control—picking up where regulation had failed—was beginning to stir.

For the federal government, it was a time of widely mixed efforts. On the one hand, despite the obvious dangers, the Kennedy Administration lifted old regulations on private power companies and allowed them to work more closely, bringing about new mergers and consolidations of smaller systems. On the other hand, the administration encouraged construction of new dams and nuclear reactors owned by public systems and development of federal transmission lines.

To undermine popular support for the public projects, a new wave of propaganda was mounted by Edison Electric Institute (EEI). Ads appearing in major publications such as *The Saturday Evening Post* and *Readers' Digest* stressed Cold War themes and took dramatic stances against public power. One full-page ad in the September 1962 issue of *Atlantic Monthly* featured the picture of a despairing young man behind a stone wall topped with barbed wire. A military figure stood anonymously in the background. "How Is Freedom Lost?" the caption asks. The answer was right out of the McCarthy era: "Dangers that grow within our borders can string barbed wire around our freedoms as tightly as dangers that come from abroad. But they aren't as easy to see. Some of us are hardly aware of the threat that grows within—the expansion of government in business . . . when government owns business, it has in its hands both political and economic power. . . . Isn't it time to call a halt to the expansion of government-in-business?" It was signed by "Investor-Owned Electric Light Companies . . . more than 300 across the nation."

Kennedy called the ads "ugly." In their 1967 book, *Overcharge,* Senator Lee Metcalf and Vic Reinemer would point out nearly $1.8 billion in overcharges to consumers for "phantom taxes"—taxes collected as part of bills but not paid to the federal government. They also went behind the scenes of the advertising campaign to reveal a much broader effort to undermine the concept of federal dams and local public power systems. Metcalf and Reinemer showed that leading power companies were helping to fund programs of the John Birch Society, which tried to equate public power and rural co-ops with communism. Twenty power companies, including Pennsylvania Power and Light, Detroit Edison, Dayton Power and Light, Connecticut Light and Power, and Montana Power Company, contributed to the Birch Society's American Economic Foundation (AEF) program in 1963. Fred G. Clark, chairman of the program, is quoted by Metcalf and Reinemer as saying, "More than 50 privately owned power companies have—in one way or another—recognized its [the program's] importance and done something about it." In addition to publishing pamphlets and disseminating literature against public power systems, the program also issued films and educational material for schools.

Between 1943 and 1965 material from the AEF had been used in economic training programs for 3.5 million workers in 2,000 corporations, 171 teachers' institutes, and workshops for primary and secondary school teachers. AEF films were permanently placed in more than 7,000 schools in 41 states. In 1961 Fred G. Clark told an audience of electric utility officials how the AEF and a power company had worked together in Pennsylvania and "within 18 months, the coverage of the public and parochial high schools in eastern Pennsylvania had almost reached the saturation point."

At the same time these events were evolving, the electrical manufacturers embarked on a propaganda campaign to encourage the vision of an all-electric future. In 1961, the electric industry planned to spend $53 million on a "Live Better Electrically" advertising campaign. Part of the effort was aimed at encouraging gas users to switch to electricity. All-electric kitchens and electric heating were promoted as symbols of personal prosperity and progress. The "Gold Medallion" all-electric home program would include kickbacks to builders who installed electric rather than conventional heating systems, a practice that later became widespread. Mason City, Washington, was promoted as exclusively using electric heating. And in the future, the concept of the all-electric city was the corporate vision of the future promoted by Disneyworld.

Both public systems and private systems, locked in their competition and the larger rationale to build and grow or be swallowed by the larger private companies, urged consumers to use more electricity. Every new gadget on the market was electric. And the new models of old standby appliances used more power than ever before. From a perspective of energy efficiency, such uses of electricity promoted inflated consumption and a massive waste of the nation's resources.

There was no better example of the waste of electricity being engineered than in the refrigerator—the major household electric appliance. A Massachusetts Institute of Technology study found that refrigerator models marketed between the years 1925 and 1950 had a power demand of 6 watts per cubic foot; those marketed between 1950 and 1965 had a power demand of 10 watts per cubic foot; and a 1972 model had a demand of 14 watts per cubic foot. This increase in consumption was above and beyond any new features such as defrosters and ice makers. The same kind of inefficiency was also engineered into other home appliances, lighting, and industrial motors. Companies making these appliances, such as GE and Westinghouse, were, of course, also making the equipment for the generating plants. Boosting demand on one side of the electrical outlet was good business for boosting the need for equipment manufacturing on the other. Consumers were unknowingly caught in the middle of this squeeze.

The "Live Better Electrically" advertising and the political propaganda had its effect. *Electrical World* magazine, a publication boosting the interests of private power companies, published a survey in 1963 that revealed a steady shift against public power companies was taking place. In 1943, 56 percent of the people surveyed approved of public power. By 1963 only 37 percent supported the idea. Conversely private power's approval had climbed 17 points to a 48 percent approval rating. In regard to nuclear power, the most dramatic changes in opinion had come since 1961, shifting from support for government ownership of reactors to a 36 to 39 percent split favoring private ownership.

At the opening of the publicly run Hanford electric plant in September 1963, the world's largest nuclear-powered facility at the time, Kennedy again expressed hope for nuclear power from both public and private sources. Private power officials responded with outrage and intensified their advertising program and their political action at the federal level. Kinsey Roberts, a member of the Business-Industry Political Action Committee and head of a Washington state power company, addressed a convention of the Rocky Mountain League shortly after the Hanford plant opened. "We in business in this country," he said, "are being badly hurt in Congress by socialists and do-gooders that we are partly responsible for sending there, and up to date we have been doing little about it financially or otherwise." Calling up the old ghost of "creeping socialism," he said public power had to be defeated. "If we don't win this fight, someone else will be next. If profit is evil for us, then it is equally evil for any other business. . . ." With the 1964 elections a year off, he asked his audience and representatives of other industries to concentrate their efforts on the federal government.

Before the elections, however, there were other tragic events ahead.

□ □ □

The assassination of John F. Kennedy in November 1963 marked an enduring turning point in history and the consciousness of the American people. Following the assassination, human and technological fallibility seemed more perceptible, limits to the New Frontier more tangible, and the sense of contradiction keener. On his move into the Oval Office, President Lyndon B. Johnson initially picked up Kennedy's power policies in his Great Society program, extending the dream of prosperity through boundless energy consumption. The darker side of these policies would continue to emerge, however.

Private power companies marked Johnson as a protégé of Sam Rayburn, a congressman who had led New Deal reform of the industry in the 1930s. They feared he was perhaps even more sympathetic to public power interests than Kennedy had been. As a member of the first Joint Atomic Energy Committee in 1947 and as a supporter of the Hells Canyon proposal and a number of other public power ventures, Johnson was seen as a believer in rural cooperatives and a threat to private domination of the industry. Even his rural roots were in co-op systems. In 1964 Johnson told a meeting of rural cooperative members, "Electricity where I live on the ranch comes from the Pedernales Electric Cooperative, one of the first organized in this country." Early photographs of the Pedernales Cooperative show a young Lyndon Johnson standing on the porch of a weathered wooden building with other co-op members. Never self-conscious about his achievements, he joked, "I had a little to do with it."

Despite these roots, Johnson had long operated in the pragmatic world of Washington politics. He had unexpected shifts in store for national power policies.

In the presidential campaign of 1964 it was no surprise that industry leaders supported Senator Barry Goldwater of Arizona, who favored dismantling the federal power program. Goldwater had long been cultivating the electric industry for support. In 1959 he had paid tribute to the influence of the private power companies at the EEI convention: "You operate in every state in the Union except Nebraska," he said. "I know of no other industry with this same advantage. You have responsible representatives in almost every Congressional district who are, or should be, well acquainted with local officials and opinion leaders." He suggested to the utility leaders that they "select a top official to handle [politics] as his primary responsibility . . . since politics is one of the most important activities that confronts you." Unlike Eisenhower's secret intent to get rid of the TVA, Goldwater also openly admitted that he was "quite serious about selling the Tennessee Valley Authority to private enterprise." His attitudes made him the favorite candidate of private power leaders and right-wing conservatives. In Florida for example, a pro-Goldwater film, *The Welfare State,* was shown in several communities under the sponsorship of Tampa Electric and the Florida Power Corporation. The narrator for the film was a well-known Hollywood actor named Ronald Reagan, who worked part-time for nearly a decade as a public relations man for GE.

Johnson did his best to build ties with private power companies. Just a month after Kennedy's assassination, in December 1963, the nuclear industry claimed it had achieved an economic breakthrough. Central New Jersey Power and Light announced it would build a power reactor on Oyster Creek, near the coastal town of Toms River. The Oyster Creek reactor was hailed as the first chosen for "economic reasons." Actually, GE offered the plant and others like it in a new marketing program, which promised fixed costs and on-time delivery. All the power company had to do was "turn the key" to open the plant door and begin operation. Westinghouse started a similar program.

Both companies hoped that losses from the first few plants would be made up in later sales. At whatever price, the fixed costs would also help to lift the shadow of the bid-rigging scandals and inflated prices from their corporate shoulders. Of the thirteen "turnkey" plants that would also literally open the door for the manufacturers in the mid-1960s, it was estimated that GE and Westinghouse lost $1 billion. But for power companies the offer of tax subsidies, insurance subsidies, and fuel subsidies from the federal government, combined with the fixed-cost scheme, was a deal hard to turn down.

Johnson linked this "economic breakthrough" with the old dream of abundance from electricity in a commencement speech at Holy Cross College in June 1964. "In the past several months," he said, "we have achieved an economic breakthrough in the use of large-scale reactors for commercial power. And as a result of this rapid progress we are years ahead of our planned progress. . . . We now can join knowledge to faith and science to belief to realize in our time the ancient hope of a world which is a fit home for man."

In a significant move indicating his intent to support acceleration of private development of nuclear power, Johnson signed a far-reaching nuclear law in August 1964, just three months before the election. The Private Ownership of Special Nuclear Materials Act gave industry the right to own fissionable materials for the first time. No longer would there be leasing arrangements with the federal government for enriched uranium fuel. A full-fledged private nuclear industry became a possibility for the first time in eighteen years.

The acceleration of the nuclear program came rapidly. Seven months after his election and less than a year after signing the private ownership act, Johnson discussed progress of the nation's atomic energy program with Glenn Seaborg. He would later note, "We discussed the progress report on nuclear power and the nuclear power plants that are being selected by American utilities because of economic considerations alone. About twenty utilities are considering such large plants and several are on the verge of announcing their orders. The commission estimates 5,000 megawatts of nuclear generating capacity by 1970 and 70,000 by 1980; seems remarkable." The figures were indeed remarkable, considering that in his report to President Kennedy just three years before, Seaborg had predicted only 40,000 megawatts by 1980. Even more remarkable was the fact that within a few years those estimated figures would double again. The first full rush for commercial nuclear plants was on.

The surge of nuclear construction plans led to an increase in popular opposition as more citizens were confronted with the idea of nuclear power plants rising up in the hills or shorelines of their towns. Behind the scenes an update of the 1957 Brookhaven report (*WASH-740*) raised the estimates on the impact of an accident at a nuclear reactor. Instead of the 3,400 deaths projected in 1957, the new report said a meltdown would result in 27,000 deaths. Estimates on people injured rose from 43,000 to 73,000. Assessment of possible property damage increased from $7 billion to $17 billion—leaving citizens and the federal government to absorb risks far beyond the insurance limitation of $560 million set in the Price-Anderson Act. The primary reason the damage estimates had increased was that the new reactors were two and three times the size of the first commercial reactors. It had been hoped that the data would show a lowering of the earlier Brookhaven statistics, which pronuclear speakers said citizen opponents everywhere had memorized. When industry officials became aware of the new figures, they worked to have the report suppressed, and succeeded. It would not be until much later that the public found out what industry and government officials already knew.

Objective insiders, however, began to change their stance on nuclear power. *Popular Science* magazine, which had naively promoted Con Ed's Indian Point One plant a decade before, now expressed grave doubt about nuclear plants. The June 1965 issue featured a cover article with the title "Is Atomic Industry Risking Your Life?" pasted over a graphic of a community being irradiated by a nuclear power plant in the background. A subheading read: "Don't look now, but there may be a nuclear plant near you. Here's the score on atomic accidents. . . ."

In addition to citing the 1957 Brookhaven report, the Windscale accident in Great Britain, and the deaths of seven people irradiated while employed by the Atomic Energy Commission, the article stated: "Power reactors are still largely experimental, and have been plagued by minor operating problems. . . . The growth of the industry and continuing pressure for power reactors in populated areas increase the likelihood of more and bigger accidents. It's also possible to question whether the industry is adequately regulated."

For the first time significant questions about the rushed commercialization of the technology and the industry orientation of the AEC were getting into the open. The growing doubt would not slow the momentum of development at this stage, but realization of the technical problems and resulting environmental and economic impacts continued to spread.

A near catastrophic accident at the Fermi plant in October 1966 would be passed off almost without mention at first. In the Detroit *Free Press* on the following day an article on an unexplained shutdown at the plant was sandwiched into 4 column inches on page 3 between a photo of the homecoming queen for Central Michigan State and a blurb on fall color in Michigan. A year later, as the shutdown extended into a curious silence, a news reporter discovered the extent of the accident and how it had been hushed up. This information heightened fears of nuclear opponents. For the most part, however, the general public was more concerned with the highly visible disasters from the rapid spread of interconnections and power pooling.

In January 1965, a blackout caused by a loose connection in a relay at a South Dakota power plant had rolled outward into Iowa and five other states in the Midwest. The sprawling size of the disaster was in part a result of the increased interconnections among the region's power companies. More than 2 million people in the heart of the midwestern winter were affected. But the event was seen as an aberration unlikely to occur again.

In the Northeast, ten months later, what would become known as the Northeast blackout—one of those events that people use to mark their personal histories—struck a six-state area and parts of two Canadian provinces. It began suddenly in the early evening of November 9 on the New York–Ontario border at the eastern end of Lake Erie.

At 5:16 P.M. 1,500 megawatts of power, equal to the output of two large generating plants, was going over five transmission lines strung northward toward Toronto. For some reason a backup relay malfunctioned, and one of the lines short-circuited. Power surged to the other four lines, which then overloaded and also went down. With nowhere to go, the power reversed its flow, instantly pouring back over the New York border. The line that paralleled the border flooded with power. Less than four seconds later, the sudden flood of power on that line overloaded the only other line into Canada, two hundred miles to the east at Massena. After it went down, power backed up from the border into New York and lines short-circuited at seven other locations. Traveling at the speed of light, the disruption came in a huge wave rippling outward toward the south and east and west over interconnected power networks in a widening circle. Overloaded power lines and generating plants went down like dominoes as the surges of power rose and fell over a region covering 80,000 square miles. From beginning to end, the process of shutdown took twelve minutes.

Broadcasters in a traffic helicopter over Boston's Southeast Expressway watched in disbelief as the city and surrounding suburbs went dark for as far as they could see. In the homes below, televisions went blank, refrigerators stopped, electric lights went out. Two hundred and fifty miles south in New York City, more than 10,000 rush-hour commuters were suddenly stranded in subway tunnels. Radios went dead, and state civil defense officials would later assure transistor radio listeners who looked out over a darkened landscape that "it was not the work of an enemy attack." While in some areas power companies shut down their regional interconnections and experienced a blackout lasting only fifteen minutes, many of the 30 million people affected were without electricity for up to thirteen hours.

It was the premier power catastrophe. Again came the claims that it was a freak accident that would never be repeated. But a similar power failure would hit the pool for Pennsylvania, New Jersey, Maryland, and Delaware in January 1967, and a few months later another would hit a region of the Northwest from La Grande, Oregon, across Idaho to Salt Lake City. The phenomenon, along with the interconnections and power pooling, was here to stay. Records the FPC began to keep in 1967 showed fifty-two outages between January 15 and July 12 of that year. About half of the equipment failures were due to weather conditions. While small blackouts had always been a part of the industry, these larger "cascading blackouts" brought on by the rapid build-up of regional power networks would continue as a serious problem, signaling the increasing fragility of the transmission systems as they grew more centralized.

Following the first major blackouts, there was an attempt in Congress to pass legislation to require more reliability in the transmission of power. The Johnson Administration called for the formation of regional councils that would give smaller systems, including rural cooperatives and municipal systems without generation, a voice in determining regional power supply planning. Determined to keep their doors closed and maintain private domination of transmission and generation of power, the private companies stopped the bill by offering to increase reliability with voluntary innovations. Eventually, in 1968 nine regional power councils gathered into the North American Electric Reliability Council to oversee power planning in their respective regions. Private power interests dominated the councils. By the mid-1970s nearly all of the 3,600 private, public, rural co-op, and federal power systems had locked the nation's 340,000 miles of overhead transmission lines into twenty-seven regional power pools.

The rationale for the interconnections, predating Kennedy's encouragement, was to keep costs down and prevent blackouts. In some cases this worked, and the concept in its pure form proved to be a noble one if checked by local control, regulatory oversight, and adequate coordination. For the most part, however, the pools and interconnections served as another avenue of private domination of the power industry. As organized in the national council, the private power companies' ability to influence regulators and growth planning was almost unassailable.

In their 1971 book, *Centralized Power,* Marc Messing and his coauthors cited analyses showing that coordination between the systems "was not substantially greater in 1970 than it was in 1963." Instead of creating the need to build fewer plants, power pooling and planning by regional councils dominated by the private companies resulted in justifying the building of more plants than were needed and ever higher surpluses of electricity. By the early 1980s these surpluses and overbuilding would present huge problems and threaten to collapse nuclear construction and bankrupt a half dozen large power companies.

"During the period in which power pool agreements have proliferated," the coauthors wrote, "reserve margins have increased almost 300 percent. . . . Large scale plants have proliferated, but purported economies of scale have not been realized. Power pools have proliferated, but utility coordination has remained limited. . . . Regional planning has proliferated but has only made the planning less accessible by state and local governments."

Power company planning conducted under the auspices of the power pools and regional reliability councils was also less open to public scrutiny. The private domination of the pools brought back the pressures of the holding companies in the 1920s. The impact on public power was illustrated by a situation in the Big Eleven Power Loop in New England, forerunner of the New England Power Pool. From 1965 through 1968, public power systems in Massachusetts faced increased pressures to sell their systems to the private companies that dominated the regional pool. In September 1967, for example, the town of Westfield was presented with an offer from Western Massachusetts Electric Company to "rent" the municipal system to the private company for $450,000 per year. Under this arrangement, the private company could use the town's distribution system and avoid paying taxes. At the time Westfield was buying most of its power through wholesale contracts from Western Massachusetts, part of Northeast Utilities holding company, which played a dominant role in the regional power pool. When the town refused the deal, its power supply was threatened. The scheme was called a "squeeze" and a "conspiracy" by an attorney for Westfield. During this same period the towns of Hull, Holden, Holyoke, and Taunton faced similar pressures. Fortunately this "squeeze" was stopped by regulatory agencies.

In the background, the power pooling and joint efforts to build nuclear plants had brought about a new wave of centralization. In New England, the Hartford Electric Company, Connecticut Light and Power, and Western Massachusetts Electric Company had gained approval in 1967 to form Northeast Utilities. The following year another proposal surfaced for Boston Edison Company, New England Electric System, and Eastern Utility Associates to merge into a new super holding company called Eastern Electric Energy System. The new company proposed to cover most of eastern and central Massachusetts as well as parts of Rhode Island and New Hampshire. Nationwide there was a huge surge toward this corporate centralization, reminiscent of the 1920s. In less than four years, twenty-four proposals for corporate mergers and affiliations among private power companies were put before the federal government.

The most ambitious proposal envisioned the combination of eight companies in Ohio, Pennsylvania, and northern Kentucky. This would have created the largest holding company in the nation: Cincinnati Gas and Electric, Cleveland Electric Illuminating Company, Dayton Power and Light, Ohio Edison, Pennsylvania Power Company, Toledo Edison, and Union Heat, Light, and Power. In the Midwest, American Electric Power absorbed Michigan Gas and Electric and proposed taking over Southern Ohio Electric. Iowa's two largest power companies, Iowa-Illinois Gas and Electric and Iowa Power and Light, would combine into Iowa Public

Service. Further south, Kansas Power and Light, Kansas Gas and Electric, and Empire District Electric Company began working toward an affiliation in 1968. To the north, Minnesota Power and Light and Northern States Power proposed combining as did Montana Dakota Utilities and Otter Tail Power Company.

This surge toward centralization resulted in an increasing loss of regulatory control over the power companies at the state level and anticompetitive tactics against public systems, such as what was occurring in Westfield. It also heralded something more. In testimony against the Eastern Electric Energy System creation, a Massachusetts attorney charged:

> This is the first major step toward placing the entire New England Electric power industry under the control of a single corporate entity. . . . Vermont Yankee's president testified before the Vermont Public Service Board that it has for some time been the industry's eventual goal to consolidate the New England power industry under the control of a single corporate entity. Unless the boycott against municipal utilities is first broken, such consolidation will make absolute private industry's ability to enforce its boycott.

Plans for centralizing power under a single corporate entity were not confined to New England alone; such plans were evident for other regions as well. In October 1967, American Electric Power holding company President Donald C. Cook told a National Power Conference in Washington, D.C., that the majority of the nation's electric systems should be eliminated because they are "obsolete, wasteful and an expensive anachronism." In a throwback to the Insull era, Cook predicted that within twenty-five years the nation would be served by a dozen or so giant electric systems. While many scoffed at the idea, the federal government began to use such a scenario in discussions about the future.

□ □ □

By 1968 the realities of inflation and public awareness of environmental damage began to tighten around the electric industry. Huge clouds of sulfurous smoke pouring from the stacks of coal-burning plants and the pollution of streams and rivers by electric power plants became targets of citizen anger. In the background, widening protests over the Vietnam War generated mistrust of the government and industry. Opposition to the modern-day "social Darwinism" of the giant corporations began to spread as people attempted to regain some sense of control over their communities and environment.

In February, popular suspicion about dangers from radioactive waste flared up when a citizen group from Rochester, New York, revealed highly contaminated drainage water from the West Valley reprocessing plant was leaching into Buttermilk Creek. One member of the group had gone under the security fence to collect samples inside the site. The group claimed that radioactive strontium 90 in water found in the drainage outlet was 30,000 times the allowable limit. A controversy involving state and federal agencies ensued. Security was tightened, and some improvements were ultimately made, but Fuel Services, Inc., which operated the facility, denied that the contamination was of a serious nature. Amid charges of more leaks and hundreds of exposed workers, the facility was shut down in 1972. Some 600,000 gallons of radioactive waste was left abandoned in temporary storage tanks.

As inflation and the growing controversy over nuclear power began to take hold, the nuclear industry seemed to hit a brick wall. In 1968 annual orders for new reactors dropped in half to sixteen. In 1969 they would drop further to seven.

Richard Nixon, the politically savvy Republican presidential candidate, picked up the grass-roots anger over environmental damage. Perhaps to divert some attention from the issue of the Vietnam War, he called the environment "the issue of the decade." Shortly after taking office in 1969 he attempted to calm concerns about nuclear power and the environment with passage of the National Environmental Policy (NEP) Act. The act required an environmental impact statement from federal agencies licensing or funding construction projects. In providing public hearings and a formal process for protest on the effects of a project, it was viewed as the most important single piece of environmental legislation ever enacted.

Other news and events also added to the momentum of the environmental issue. After six years of study on the effects of low-level radiation, two Atomic Energy Commission scientists, Dr. John Gofman and Dr. Arthur Tamplin, found a potential for widespread danger from operating nuclear plants. Effects of low-level radiation, they said, had been vastly underestimated in the past. If average exposure to the United States population was maintained at the current limits, the effects would accumulate to an excess of 32,000 fatal cancer and leukemia cases per year. Beyond the immediate impact, they claimed there was also the danger of far-reaching genetic damage. They recommended that exposure limits be reduced by a factor of 10. Industry leaders and AEC commissioners were stunned by the report and worked to discredit the findings. Both scientists came under extreme pressure, and Gofman resigned from the

AEC, charging that he had been harassed into leaving. The information he and Tamplin released, however, provided ammunition to the growing environmental opposition to nuclear power.

To gather the growing concern, environmentalist organizations sponsored a nationwide Earthday celebration in April 1970. The event was an eclectic coordination of local environmental protests and "teach-ins" in universities, state legislatures, and town squares. In New York City, 1,000 youthful supporters gathered at the corner of Wall and Broad streets to hear Senator Jacob Javits warn that environmental organizing should not distract from social problems of the poor. Looking out at the offices of J. P. Morgan and the Stock Exchange, other speakers charged that the quality of life in America was "being destroyed by the people who control the country." In Detroit that same day, forty women picketed the Great Lakes Steel Corporation to protest industrial discharges into the Detroit River. This was one of dozens of local protests taking place around the country. Out of all this activity would come momentum for the Clean Air Act of 1970 and the Clean Water Act of 1972, mandating the states to set pollution regulations and enforce them.

Despite the passage of these laws, government officials and industry executives still clung to their dream of a society based on unbounded supplies of energy. A campaign was mounted to get the nuclear industry back on track. Glenn Seaborg, chairman of the AEC, published a *Looking Backward*–style article in the *Bulletin of Atomic Scientists* in June 1970. From the vantage point of 1995, Seaborg related how the debates over nuclear power had subsided beginning with a SALT agreement and slowdown and halt of the arms race. He said a substantial amount of evidence was presented to the public that "no harmful biological effects would be caused by the radioactive effluents released from nuclear plants." More pointedly he wrote:

> The massive nuclear accidents predicted by alarmists at the height of the anti-nuclear campaign never occurred. At times nature did her best to knock out nuclear installations. In the Far East a nuclear station sustained some damage from the wind and flooding of a typhoon. And a nuclear plant on the West Coast of the United States felt the effects of an earthquake. Although there was costly damage to both plants, their containment systems were not breached, and in neither case was the public exposed to any serious radiation hazard.

By 1990, Seaborg wrote, "Public confidence in nuclear power had risen to the point where there were few eyebrows raised when a midtown reactor was proposed for a new city being planned, because many such reactors had been operating in urban sites." Seaborg's vision was one of producing a "steady state" civilization in which nuclear power would help a controlled world population live in harmony with nature.

Just as investment banker Lewis Strauss delivered nuclear technology to private interests, Seaborg dedicated himself to accelerating its development to fulfill the naive dreams of atomic visionaries of the early 1950s. Throughout 1970, the AEC promoted the concept of "abundant nuclear energy." Despite the steady escalation in the cost of plants, commissioners claimed that nuclear power was the technology that would bring about a clean environment and a decline in the cost of electricity. The Atomic Industrial Forum and EEI got behind the AEC in an effort to bring back the market for reactors. Advertising stressed not only all-electric homes, but "infinite energy" for families and "all-electric cities" for the future. Their success would be nothing less than spectacular. In 1970, fourteen reactors were ordered; twenty-one in 1971; thirty-eight in 1972; and a high of forty-one in 1973, before the industry was hit with the Arab oil embargo and the shock wave of rising costs and consumers voluntarily cutting back on inflated use of electricity.

While the power companies renewed their interest in nuclear power, the promotional campaigns didn't convince the public. As announcements and construction of some 106 new reactors in towns throughout twenty-nine states took shape, popular concern broadened. Fed by steadily rising costs, an antiestablishment attitude engendered by Vietnam War protests, and increasing awareness of environmental hazards, mistrust of the electric power companies deepened.

In December 1970, scientists who had been relatively silent confronted Seaborg. The group, loosely organized as Scientists and Engineers for Social Action, appeared at a meeting of the American Academy for the Advancement of Science, at which Seaborg was scheduled to speak. Seaborg left the meeting without speaking. A statement he issued to the press said, "I think we must realize that science is suffering today from a kind of dislocation and disunity brought on by its own success. That is, it has fostered changes in our society faster and with more impact than our social and political understanding can absorb." It was a curious twist on statements Einstein had made twenty-five and thirty-five years before. While Einstein had urged that science rely on the public to decide what risks would be taken, Seaborg implied that the public rely on scientists.

Extending Seaborg's rationale, many power industry officials believed that they were the ones who should make the commercial decisions for science and determine what environmental and economic risks the public should take. They strongly resisted the new Clean Air Act, which appeared to them in the form of increased costs and delays in construction programs. The backbone of the electric power industry at the time was in 665 fossil fuel plants (coal, oil, and natural gas) which together provided 80 percent of the electric power in the United States. Coal plants, the nation's worst source of air pollution, were a chief target of environmentalists. President Nixon and industry officials believed that nuclear power was "safe, clean, and reliable" and offered the only alternative to the pollution caused by coal-fired plants.

The public, which had been left out of the decision making, began to demand a greater voice. Voters in Eugene, Oregon, halted construction of a nuclear reactor by passing a four-year moratorium to study potential hazards from the plant. On the opposite coast, Long Island residents went into a local elementary school auditorium for what would become nearly a year of environmental hearings on the license for the Shoreham nuclear power plant. And in Maryland, the proposed Calvert Cliffs plant became a test case over whether the AEC would comply with the environmental impact statement required by the NEP Act. Commission attorneys argued that environmental considerations were outside the commission's jurisdiction, but the courts upheld environmental groups' demands and required the commission to make an independent review and evaluation of all effects at every point in its licensing process. The ruling threw the status of more than fifty nuclear projects into a temporary limbo. Amid much criticism the commission hastily processed the paperwork on the plants to comply with the letter if not the spirit of the law. The events surrounding the Calvert Cliffs decision seriously embarrassed the commission by emphasizing its conflicting mandate both to regulate and promote nuclear power. Wheels were set in motion that would split the agency in a few short years. For environmentalists, the cases pointed to the fact that alternatives for electric power were needed.

Environmentalists wanted clean energy, but many of the traditional environmental organizations were undecided about the use of nuclear power. In 1971, the Sierra Club published *Energy,* a book in which the authors claimed that if the 1,000 reactors projected were in operation by 1990, it would mean the nation "could then expect one accident in the U.S. every year." To debate this danger and possible courses for the future, the Sierra Club organized a meeting of four-hundred environmentalists, regulatory officials, and industry representatives in Vermont in January 1971. The purpose was to try to formulate an environmentally responsible energy policy. What resulted were recommendations for energy conservation and energy efficiency; improving appliances; reducing industrial use of electricity without sacrificing jobs; changing tax structures and rate structures; and tightening regulation of the industry.

One speaker at the conference, Julius Hobson, a professor at American University, stirred controversy by pointing to the "myth of free enterprise" in the electric power industry. Hobson said that the industry existed in a form of "corporate socialism" or "private socialism" that was determining the future of the nation. "I do not believe that what is good for the utility companies is good for the United States," he said and called for a national takeover of the electric industry. The conference's final recommendation to the Sierra Club members urged a "moratorium on nuclear power plants as presently designed and installed." Not quite as strong as Hobson's appeal, but a significant stance for the Sierra Club to take.

A vastly different perception dominated the industry and federal government. Industry executives believed that alternative technologies (such as solar, wind, and geothermal energies) held promise for the future but were "impractical" in the short term. There was a more fundamental problem as well. Such decentralized power sources were not technologies that could be utilized for increasing centralization of the electric power industry and for laying the base for greater corporate expansion in the future. In a harsh way, Hobson had struck the central issue of the energy debate and the one that had existed since the beginning of the electric industry: Who really controlled the decisions over choices of technology? Was the federal government promoting what industry and Wall Street wanted or what the public wanted? The answer, of course, was tied to an increasingly central part of the economy and the risks individual communities would face.

There was no question of where President Nixon stood, or where his views originated. Under Project Independence he called for 50 percent of the nation's electricity to come from 1,000 nuclear plants by the year 2000. In promoting this centralized vision of the future he predicted in 1972:

> The most profound effect on electric utilities through the year 2000 will probably be a completely new restructuring of the industry. Emerging will be a small number of super G&T [generating and transmission] organizations. Utilities as we know them today will be relegated to the distribution function. Donald C. Cook, chairman of American Electric Power, five years

> ago was the first to suggest that the regional nature of power generation requirements ultimately could lead to the emergence of only 15 to 20 super G&T utilities serving the demands of the entire U.S., with distribution being provided by the utility organizations as we know them . . . increased demand for electrical energy will lead through normal progression to pool arrangements becoming more and more formal . . . some of the 15 to 20 separate G&T utilities will result, owned in part by those utilities performing the distribution function, but under the direct control of boards composed of representatives of government, consumers, and the system operators.

Aside from being in agreement with Donald Cook, whose holding company sprawled over parts of seven states and would undoubtedly be one of the "super G&Ts," Nixon said the concept also matched with the results of a "special management think-tank exercise investigating the organization of the electric utility industry into 2000 and beyond, sponsored by the Edison Electric Institute." He said this group envisioned a change in the ownership and control of electric utility organizations to "quasi-public corporations, similar to that of the Communications Satellite Corporation (Comsat)." Essentially this meant that federal funds would be used for the private development of these "super G&Ts."

Earl Butz, Nixon's controversial Secretary of Agriculture, issued a similar statement concerning the future direction of the nation's rural electric cooperatives for which he had oversight. "The prophets of doom," Butz said, referring to environmentalists and others advocating programs for conservation, "tell us the supply of electric energy is running out. . . . They warn that a nation of wastrels, using electricity to power toothbrushes, comb dogs, feed fish, and dry fingernail polish is on a collision course with a colossal power shortage—a catastrophe that will catch up with us between now and the turn of the century." He brushed aside these complaints and the cost of the larger institutionalized waste they symbolized by saying, "We live in an energy-oriented society. That's the way it should be." Like President Nixon, Butz said the answer was in "a nationwide network of transmission facilities, and a correlated grid of public and private power systems. . . ."

The similarity to the 1920s, in the rise of fifteen to twenty large companies to provide the nation's power, was clearly visible. Although there was an indication that government officials and consumers would also be present among their directors, the private companies were very likely to dominate the politics of the organizations. One federal official said, "The new mandate for enlightened utilities, as we see it, will be to take a more active role in political planning bodies, development authorities, and urban renewal projects. . . . Utilities cannot be expected to dive into this directly, since it is a sensitive area politically, and previously outside their scope of activity." He said that along with investment houses the utilities will " 'get involved' indirectly at first when attempting to optimize load projections, predict growth and accommodate commercial and industrial development."

Buried in that involvement is the subtle control that public officials had feared since the nineteenth century. Now the federal government was advocating it and attempting to promote the centralized visions of the future that would help bring it about.

Faced with threats of fuel shortages and increasing costs and debilitated by the long battles, even the municipal power systems sought to work together with private power companies to solve technological problems leading toward this centralized vision. While conflicts remained sharp in some regions, public systems joined with private power companies in creating the Electric Power Research Institute (EPRI), a research and development organization designed to come up with solutions to problems of centralized technology. EPRI would ensure that public and private power systems shared common visions of a future powered by breeder reactors and fusion reactors. Just as public system participation in the North American Electric Reliability Council captured their transmission and growth planning, EPRI's private power orientation would harness their technological vision to the centralized plans of private power. The early mandates of local public power systems were now obscured. Despite their potential to bring citizen control over electricity, the options for many public systems were almost completely in the hold of private power.

The most prominent symbol of what would soon become widespread joint ventures of public and private systems occurred between Insull's original company, Commonwealth Edison of Chicago, and the TVA. In answer to a call by President Nixon that the nation have a commercial breeder reactor operational by 1980, TVA and Commonwealth Edison signed a memorandum of understanding in August 1972 as principal participants in the Clinch River Breeder Reactor. The project engendered a long battle with environmentalists and consumer advocates. The Clinch River project would eventually be stopped by environmental and consumer protest in 1983 after nearly $1 billion had been spent and ground breaking had just been completed.

There were much worse financial disasters brewing for the industry. In June 1972, *Business Week* ran a cover story predicting a crisis in the electric power industry. Huge amounts of capital were needed by the industry to build new plants and refinance the enormous bond issues floated at the end of World War II, which would be coming due over the next several years. At the same time the companies faced tightening environmental restrictions and state regulators attempting to hold rates and inflation down. A series of brownouts and blackouts in major eastern cities had dramatized the dilemma. It was obvious that the industry, preoccupied with public versus private conflicts and sluggish management, had suffered gross neglect during the 1950s and 1960s. One analyst, Ted O'Glove of the industrial research firm Coenon and Company, told *Business Week,* "I think the blame lies with their managements. When times were good, they didn't campaign for higher rates because they were afraid of rocking the boat. The regulatory agencies are to blame, too, because they haven't reacted to the problem."

In the same article, American Electric Power Company's Donald Cook observed, "The days when things were easy are gone." Cook was indeed right. Consumer groups were now regularly intervening before regulatory commissions on issues of rate increases and environmental complaints, and there would be a tough fight to get the money the power companies wanted to pay off their debts and launch new building projects. But this crisis for cash was obscured by a much larger event—the Arab oil embargo.

Trouble over control of the flow of Middle East oil had been brewing for some time. In 1973 the United States was receiving 30 percent of its oil imports from the Middle East. When the United States backed Israel in the Arab-Israeli conflict of 1973 by airlifting emergency arms, the Saudis embargoed all oil shipments to the United States. The long-predicted fossil fuel shortage, due to political rather than purely resource limits, hit with tremendous impact and underscored the international scope of the energy resource problem.

With American power companies dependent on oil for 17 percent of their generation (some individual companies as much as 75 percent), the embargo sent oil prices and electricity prices soaring, particularly in the Northeast. The power industry responded with wave after wave of emergency rate increase requests. State regulatory agencies became hotspots for consumer frustration as they buckled to industry's requests. As rates began to soar, the enmity for power companies, which had been growing since the early 1960s, erupted. Regulation had failed, the yardstick of public power was captured, and consumers had little institutionalized voice. In hundreds of conflicts throughout the 1970s and 1980s, citizens sought to bring back some sense of control over this runaway industrial empire and the risks it was bringing to their communities.

6. Citizen Rebellion

It is a protest against the abuses of industrial technology that poison land instead of nurturing it, that sour the air and foul the water, that devour marsh and woodland and make hazardous to health and peace of mind the cities and factories in which people live and work. It is a protest against governments—all governments—that spend billions in an endless, insane atomic arms race that consumes the cream of the world's resources and much of the brightest human talent. It is a protest against the uses of science and technology that are antihuman and antilife.

David Lilienthal
Atomic Energy: A New Start
1980

On the warm, breezy afternoon of April 30, 1977, in a dusty parking lot in southern New Hampshire, the mass movement against nuclear plants that had been stirring for a decade finally came to life. Carrying backpacks, sleeping bags, and tents, more than 2,000 antinuclear activists converged on the partially constructed Seabrook nuclear project. Overhead, police helicopters and a small plane buzzed above the trees and surrounding marsh, and warnings against trespassing crackled through bullhorns. One contingent marched along the access road from the town's main thoroughfare. Another slogged across the surrounding salt marsh. Other smaller groups were ferried across Hampton Harbor by sympathetic lobstermen and landed on the beach above what would become the mile-and-a-half-long cooling tunnels running out into the ocean. With colorful banners and American flags flying, the protestors announced that they had come to "occupy" the site until construction was stopped.

Debate had been going on over the Seabrook site since 1968. Public Service Company of New Hampshire—a small private power company that had once been part of Samuel Insull's network—at first told local residents that the company was going to construct an office building.

During the ensuing years, as plans for twin nuclear plants became known, the debate deepened, resulting in a town meeting vote against the idea. Nothing seemed to dissuade the company, however.

On August 1, 1976, following a rally of six hundred people on the town common, eighteen New Hampshire residents had walked to the site with the intention of stopping the construction and were arrested for trespassing. Three weeks later, on August 22, following a larger rally, 180 protestors returned to the bulldozed site and refused to leave. Word of the demonstrations spread.

During the winter months, leaders of the local Clamshell Alliance began planning a larger demonstration. They sent a call to environmentalists across the country to take part in the protest. Borrowing an organizing structure and principles long used by Quakers, activists were organized into "affinity groups" of eight to twenty persons and trained in the principles and tactics of nonviolence. These groups, each with its own spokesperson, became the functioning units that provided personal support and structure that turned a diverse group of people into an organized community. It was to be a national symbolic showdown with those who advocated nuclear construction. While most of the demonstrators were from New England, others came from as far away as Florida and Alaska to take part in the mass civil disobedience.

Hundreds of state police ringed the security fence around the site and watched as the occupiers set up avenues of tents and selected leaders for a coordinating council, which would develop a system of security and help run the encampment. On Sunday, May 1, Governor Meldrim Thomson moved school buses to the site and arrested 1,400 demonstrators. Refusing to pay bail or participate in the judicial process, they were held in state armories for nearly two weeks. Media coverage of the event was widespread and brought the growing debate over nuclear power into the living rooms of America. In the following months, similar alliances and mass protests already brewing emerged in communities all across the country.

The conflict over nuclear power, above all other issues, became the major dispute between an increasingly distrustful public and the electric power industry. It magnified all of the back-room dealings of the industry and brought attention to focus on the re-emergence of a determined new power industry empire made up of the Edison Electric Institute (EEI), Atomic Industrial Forum, major construction companies such as Bechtel Corporation, Exxon Nuclear, Westinghouse, General Electric, a host of major banks such as Chase Manhattan, and Wall Street brokerage firms.

Like the power empire of the 1920s, the new industrial empire coordinated its extensive political and economic clout to get what it wanted. And as in the old days, it was backed by an array of local and national politicians who received regular campaign support from the industry.

At the local level, between the time of the oil embargo of 1973 and the demonstrations at Seabrook, several courses of action had been attempted to resolve the growing conflict over nuclear construction. Until the mid-1970s the debate over nuclear power had been largely confined to state and federal hearing rooms. Dozens of small antinuclear groups —which sprang up in the late 1960s and early 1970s as announcements came out for construction of more than one hundred reactors across the country—usually focused their efforts on intervention in AEC licensing hearings. It was in those ponderous, often year-long proceedings that many citizens hoped to prove the case against nuclear power.

Lengthy interventions against the Calvert Cliffs plant in coastal Maryland, against Shoreham and Indian Point in New York, and against plants in California, Michigan, and Vermont led to a series of legal challenges and precedents that laid the groundwork for future debate over nuclear energy. In nearly all cases, the intervenors failed to halt the construction or operation of the reactors. But they did succeed in winning environmental concessions from power companies, and they gathered public attention and forced explanation of the dangers involved.

The intervention of the New England Coalition on Nuclear Pollution in the Vermont Yankee licensing hearings, for example, succeeded in pushing the AEC to require additional safety features, including the construction of cooling towers—mammoth conelike structures rising out of the New England woods—to prevent thermal pollution of the Connecticut River. This organization, which was formed by citizens and scientists from Vermont and western Massachusetts in 1971, later took part in a lawsuit brought before a federal Court of Appeals in Washington, D.C., to stop the plant from going on line in 1973. The intervenors asked for a second review of Vermont Yankee on grounds that plans for the permanent disposal of radioactive waste had been excluded from the initial environmental review. The federal Court of Appeals agreed that the reactor license was deficient because the government had not adequately considered the environmental impacts of waste disposal and fuel reprocessing. Following the decision, for almost three months, all licensing of new nuclear plants was suspended. It resumed soon after, however, when the new Nuclear Regulatory Commission (NRC) completed a hurried study of the environmental consequences of fuel reprocessing and waste management. The NRC allowed the power companies effectively to sidestep the issue of reactor wastes, which would remain dangerously radioactive for more than 100,000 years.

The clamor over nuclear safety emerging at the local level echoed in the halls of Congress. In the wake of the oil embargo in 1974, legislators took up a tumultuous reshuffling of energy agencies, including an unsuccessful attempt to divide the conflicting promotional and regulatory missions of the AEC. What finally resulted was creation of the NRC and the Energy Research and Development Administration (ERDA). But other congressional efforts to strengthen safety requirements of nuclear power plants and to provide funds for consumer and environmental intervenors were blocked by the power industry. This legislative failure further galvanized antinuclear activists, who sensed they were being cut off from the larger issues and left to fight isolated battles in their local communities.

When it became apparent that the first major debate over reactor safety on Capitol Hill was lost, Ralph Nader organized a national antinuclear conference in Washington, D.C. The 1,200 activists who attended this first Critical Mass conference in 1974 listened to a variety of speakers and attended workshops where ideas about tactics and strategy were shared and exchanged. What arose from the national conference was a call for increased citizen intervention and joint action by environmental and consumer groups fighting for utility rate reform. "A coalition of these two movements," Nader told the participants in the conference, "is essential if either is to succeed. Citizens fighting utilities as nuclear proliferators need to understand the strategy of cutting off the flow of money to utilities for nuclear expansion—by entering rate fights and stockholder and mismanagement suits."

As environmental and consumer groups began to implement this strategy in 1975–76, support for the antinuclear movement broadened to encompass more diverse segments of society, including community organizations, Native American and farmer groups, and organized labor.

Several incidents occurred that helped build the new wave of antinuclear organizing and trigger an intense period of protests between 1976 and 1979. The first incident was the disastrous fire at the TVA Brown's Ferry plant in February 1975, which severely damaged the nuclear industry's credibility. The fire was started accidentally by two electricians using a lighted candle to test for air leaks in the cable spreading room. It knocked out the main controls and safety systems of the 2,200-megawatt plant. The operators struggled for more than fourteen hours to regain control of the plant, while the NRC and TVA issued bland press releases describing the accident as a "nuclear excursion." The details of the accident were brought to the attention of millions of Americans in October when *Newsweek* published a capsulized version of an account written by David Comey, the director of environmental research for Business and Professional People in the Public Interest. Comey's account was largely based on internal reports obtained through a formal legal request, after hearing some of the details from plant employees. "This near miss," Comey wrote, "came hair-raisingly close to sending two huge Alabama reactors on their way to China in a meltdown." The public was further shocked by the publication of John C. Fuller's book, *We Almost Lost Detroit,* which appeared on bookstore shelves a short time later. The book provided details of the cover-up surrounding the harrowing accident at Detroit Edison's Enrico Fermi reactor in 1966. News of these narrowly averted disasters generated broad debate about nuclear safety and encouraged utility activists in California and other states to take the issue of nuclear safety directly to voters. Petition drives in a number of states were started to place nuclear moratorium initiatives on ballots. If passed, the initiatives would allow the construction of nuclear plants only if safety systems were proved, radioactive waste disposal processes in place, and liability limits under the Price-Anderson Act were removed.

The highly publicized defection of four people from the ranks of the industry in February 1976 shocked the public with new revelations and added to the atmosphere surrounding the initiative proposals. Three GE engineers, who resigned on February 2, told the Joint Committee on Atomic Energy that nuclear power was "so dangerous that it now threatens the very existence of life on this planet." These men, Dale Bridenbaugh, Richard Hubbard, and Gregory Minor, had managed quality assurance, performance evaluation, and advance control programs, respectively, at GE's nuclear facilities in San Jose, California. They spoke from experience, citing a number of specific design flaws in the widely used boiling-water reactors. They also expressed concern over the adequacy of quality assurances and reliability testing and the threats of human error and sabotage. The cumulative effects of flaws in design, construction, and operation of nuclear plants, they said, made a nuclear power plant accident a virtual certainty. The only questions were where and when.

The fourth defector, Robert Pollard, was a federal safety engineer for nuclear reactors at Indian Point in New York and a project manager for the NRC. He resigned because the "plants were unsafe and posed the danger of a major catastrophe." Pollard told the Joint Committee that the Indian Point nuclear station constituted an unconscionable threat to the health and safety of millions of people in metropolitan New York.

"The Indian Point plants," he said, "have been badly designed and constructed and are susceptible to accidents that could cause large-scale loss of life and other radiation injuries, such as cancer and birth defects." The magnitude of the hazards associated with the plants, he noted, "has been suppressed by the government because the release of such information might cause great public opposition to their operation."

The old battle lines between public and private interests arose as the proposals for nuclear construction moratoriums moved toward the ballot box in seven states. The new power empire responded with massive spending and slick advertising campaigns. Organizers of the California initiative, who had only small donations to work with, watched the massive countercampaigns take shape in the summer of 1976.

The industry formed citizen front-groups and spent from ten to one hundred times as much money as the proponents. A pronuclear organization called Citizens for Jobs and Energy in California, for example, put together a $2 million campaign to convince voters that the measure was a blanket ban on nuclear construction, which would cost the state jobs and money. The largest contributor to the countercampaign was Pacific Gas and Electric Company, a utility serving most of northern California, including San Francisco. This company alone poured more than $500,000 worth of manpower, service, and cash donations into the initiative fight. Other corporate contributors included Exxon Nuclear, Westinghouse, Bechtel, and more than thirty power companies from across the nation, which were eager to stop the state moratorium strategy in its tracks. The California initiative was defeated by a 2–1 vote.

To defeat Oregon's proposal, the power industry and its allies spent $900,000. More than $600,000 was spent to beat back Arizona's nuclear ballot item, and almost $500,000 went to defeat Washington State's nuclear moratorium initiative. In these three states as well as in Ohio, Colorado, and Montana, the industry used virtually the same argument as in California: the ban on nuclear construction would cost jobs, increase taxes, and lead to blackouts and high electric bills.

Frustrated by the failure of legal intervention and voter referenda, antinuclear groups all across the country were looking for a new strategy when the mass civil disobedience began in New Hampshire. Although sit-down strikes and occupation of industrial plants were traditional tactics of organized labor and the civil rights movement, they were untried for environmental issues. A Clamshell organizer, Harvey Wasserman, noted that the tactic "seemed to be our last resort. No one was winning any legal interventions, and there was no prospect of government action."

The first hint that there would be a different kind of opposition in New England came on Washington's Birthday in February 1974, when twenty-seven-year-old Sam Lovejoy, an organic farmer from Montague, Massachusetts, used a crowbar to loosen guy wires and topple a 500-foot weather tower that had been erected at the site proposed for a twin reactor near his home. After watching the tower crash, Lovejoy turned himself in to the local police, who at first refused to believe his story.

In September, Lovejoy and his supporters were in court for a widely reported trial. The case was dismissed on a technicality, but members of the jury later told reporters they would have acquitted him anyway. Debate erupted in the surrounding Pioneer Valley of western Massachusetts, and antinuclear forces succeeded in winning a 47.5 percent vote against the proposed Montague plant in a November referendum. A third of the voters also approved dismantling nearby plants operating in Rowe, Massachusetts, and Vernon, Vermont.

An individual act of civil disobedience similar to Lovejoy's occurred in New Hampshire at the Seabrook site when Ron Rieck, a twenty-four-year-old activist, climbed the 300-foot weather tower in January 1976 and remained there in freezing weather for thirty-six hours to dramatize opposition to the proposed reactor. Environmental organizations were also involved in the fight against Seabrook. Members of Save Our Shores, who defeated shipping magnate Aristotle Onassis' plan to build an oil refinery in Newington, New Hampshire, turned their attention to Public Service Company's plans to build the twin reactor at Seabrook. After reorganizing as the Seacoast Anti-pollution League, they joined with the New England Coalition on Nuclear Pollution, New Hampshire Audubon, and other intervenors to challenge the environmental soundness of the company's plan to construct the reactors. Intervenors argued that an earthquake fault (the Boston-Ottawa fault line) lay offshore; that the delicate marshland was the worst possible place to construct a reactor; that it wasn't needed to provide New Hampshire with electric power since demand was declining and the company had excess power of 19 percent; that the cooling tunnels, which would expel more than a billion gallons of superheated water into the Atlantic Ocean a mile and a half offshore, were a danger to marine life, especially the soft-shell clam beds; and that Public Service Company was too small to afford to be the major owner of the plant. The intervenors also argued that, if the plant was built and an accident occurred during the summer, it would be virtually impossible to use the town's two-lane highway to evacuate the 100,000 tourists who visited the popular beaches north and south of the plant.

During the hearings on the projects, Daniel Head, chairman of the Atomic Safety and Licensing Board, listened attentively to their concerns and arguments. In a move that many considered more than coincidence, Head accepted a new assignment before a decision was rendered. By a 2–1 vote the reconstituted board granted a conditional construction permit, pending final Environmental Protection Agency approval of the controversial cooling tunnels. Testimony from power company officials later showed that federal officials had contacted power company executives concerning the performance of Daniel Head and on the possibility of his transfer. The machinations surrounding the decision were instrumental in convincing antinuclear activists that it was time to try other tactics.

In addition to examples of civil rights and labor protests in the United States, the plan for civil disobedience at Seabrook was inspired by mass actions in France and Germany, where thousands of demonstrators had been staging protests to nuclear plant construction since 1971. The success of a citizen occupation in blocking construction of a plant at Whyl, Germany, in February 1975, where 28,000 people overran the site, was an especially vivid model. Across the United States activists frustrated with the political clout of the power industry in the regulatory process were ready for the new strategy.

By the time the 2,000 Clamshell occupiers marched onto the Seabrook site in April 1977, similar occupations were being planned for the Diablo Canyon reactor in California and for the Trojan plant in Oregon, already in operation as the nation's largest single reactor. Some protests began to link up the historical connection between nuclear power plants and nuclear weapons. On August 6, 1977, eighty-two members of Oregon's Trojan Decommissioning Alliance were arrested when they blocked the main gates of the Trojan nuclear plant and demanded an immediate shutdown and decommissioning of the reactor. A day later, ninety-two members of the Abalone Alliance were arrested when they tried to occupy the Diablo Canyon plant site in San Luis Obispo.

By the time these acts of civil disobedience, and more than a hundred others that spread in their wake, began, the nuclear industry was in deep trouble. While the protests were damaging the image of the power companies, soaring costs of construction and declining demand for power were taking a heavier toll. During 1976, power companies ordered only three new reactors compared to five in 1975 and twenty-seven in 1974. The industry was also hit by a wave of cancellations and deferrals of plants that had been ordered earlier. In the first three months of 1977, power companies in seven states canceled three reactors and deferred seven. By the close of the year, announcements of deferrals and cancellations had become routine; as many as thirty-six reactors were deferred that year alone, some for the second time.

By 1978, antinuclear sentiment had spread to farmers, labor leaders, and community activists, who began to see the broader implications of nuclear development. More than environmental pollution or consumer issues, concerns now turned to land use, the loss of jobs to an all-electric nuclear economy, and nuclear weapons proliferation. The strength of nuclear opposition intensified as these diverse constituencies added their voices to the protests. In March 1978, a local coalition of California farmers and farm workers took the issue of nuclear power to the voting booth and successfully defeated a nuclear power complex proposed for the town of Wasco. In the upper Midwest, local farmers played a large role in urging the cancellation of the Tyronne reactor planned for a site twenty miles southwest of Eau Claire, Wisconsin. Other farmers demanded the shutdown of plants at Monticello and Red Wing, Minnesota.

In the background, the stage had been set for a surge of popular support for renewable energy and conservation. The Sierra Club conference in Vermont in 1971 and follow-up meetings and discussions had gathered a political force well aware of the alternatives to nuclear power. Several major studies had emerged showing the wisdom and economic advantage in conservation and alternative energy development. One major report, *A Time to Choose,* had been released by the Ford Foundation Energy Policy Project in 1974. The report urged sweeping energy reforms and citizen participation at every level of decision making. In essence it helped to lay the basis for much of the work that would come later from Amory Lovins and other alternative energy advocates. It attacked the old assumptions of prosperity derived from boundless electric supplies directly. "The future rate of growth in GNP (Gross National Product) is not tied to energy growth rates," the report stated. "Our research shows that with implementation of the actions to conserve energy in the years ahead, GNP could grow at just under 2 percent. . . . The project also finds that it appears feasible, after 1985, to sustain growth in the economy without further increases in the annual consumption of energy." The report was widely criticized by power industry officials for its "populist approach" to energy problems.

Backed by government statistics, a Washington, D.C., group, Environmentalists for Full Employment, published reports showing that a dollar spent on solar and conservation would produce ten times more jobs than a dollar spent on nuclear power or fossil fuel development. This reasoning took root, and a dialogue on energy and jobs heated up in the labor movement. As the strategies for solar energy, which had been ignored in the early 1950s, came back to life, a number of United Steel Worker locals and Meat Cutter locals voted resolutions to oppose nuclear construction. And the International Association of Machinists, the Sheet Workers International, and the United Auto Workers took strong pro-solar stands. By May 3, 1978, when a nationwide promotion of solar and conservation technology was celebrated as Sun Day, it seemed that all of the citizen opposition might indeed create a new and better technological alternative.

The upshot of the broadening movement against nuclear power was the passage of the Public Utility Regulatory Policy Act of 1978, which promised a tremendous boost to solar and renewable energy development. If fulfilled, the law could bring about sweeping decentralization of the power industry. The private power companies resisted and stalled these changes, however, showing that control of the nation's most critical energy resources were still firmly in their hands. Nuclear construction was the most visible problem, but there were a host of other conflicts tied to expansion of the nation's electric systems.

□ □ □

Although ordinary citizens had only limited access to technical information on the dangers associated with atomic technology, they sensed something had gone terribly wrong, not just with nuclear power, but with the industrial and governmental forces that produced it. The Federal Power Commission's recommendations in 1960 to triple the nation's electric output by 1980 had fostered more than accelerated commercialization of nuclear power. There were also massive efforts to strip-mine land, build networks of extra-high-voltage power lines, and flood out communities to create new power resources.

In the early 1970s, as these projects came into being, rural residents all across America started waging a rebellion against the plans of power companies and their corporate allies to increase production and transmit wholesale power to already glutted urban markets. They resented having valuable farmland sacrificed and being relocated or used as guinea pigs to meet the needs of the electric industry. At the same time the wave of antinuclear demonstrations was going on, people in many rural communities fought industry expansion in order to protect their civil rights, their land, and their communities.

Although displacement of families in the Tennessee Valley and other areas in the 1930s and the destructive force of mining had always been a problem, what began in the late 1960s and 1970s represented a whole new scale of exploitation, pollution, and other hazards. A Library of Congress study, *Energy and Landscape,* estimated in 1976 that as much as 45 million acres of land in the United States, a land area equal to the state of Missouri, would be used for energy production by 1985. Another study found that the transmission lines planned for the nation would take up another land area the size of Delaware. Seven coal slurry pipelines, proposed to carry a gooey mixture of pulverized coal and water thousands of miles across fifteen states, would pose a far-reaching threat to water-scarce Western states in which they were to originate. Scores of new coal-fired plants were to be built and new dams were to flood out tens of thousands of acres. In many communities where these plans emerged, resistance bloomed.

At "Big Mountain," in the Four Corners area seventy miles northeast of Flagstaff, Arizona, Navajo women provided the spark for resisting government plans to forcibly remove 10,000 Navajo and 100 Hopi people from their homeland in order to make way for expanded mining operations. In 1974 Congress had passed the Navajo-Hopi Land Settlement Act to resolve a dispute over a 1.8 million acre Joint Use Area, a region the size of Rhode Island. The widely publicized agreement attempted to resolve an alleged dispute by relocation. Also thrown into the deal were national energy policy and federal development plans for the Southwest. According to many traditional Hopi and Navajo people, the land dispute had been created as a ruse to remove both Hopi and Navajo from the Joint Use Area so that energy companies could mine energy resources at Big Mountain.

The first act of the deal, along with relocation, was construction of a 400-mile-long barbed wire fence. When the fencing crew came to Big Mountain in 1977, they were confronted by Pauline Whitesinger, a forty-three-year-old Navajo widow. She demanded that they stop building the fence, but they refused to listen. Angered by an obscene remark from the crew foreman, Whitesinger knocked him to the ground and convinced the crew that they should leave her land by hitting them with a stick and throwing dirt. She and a friend, Roberta Blackgoat, later tore down a portion of the fence. Building of the fence continued, but not at Big Mountain.

Two years after the first incident, in September 1979, Katherine Smith, a sixty-year-old elder of Big Mountain and Whitesinger's half sister, was arrested for firing her rifle to stop another government fencing crew. When a crew again returned to Big Mountain in December 1980, they were once more confronted by a group of women who tore up the fence posts. The ensuing scuffle ended with the women Maced and on the ground in handcuffs.

In July 1984 another confrontation took place over attempts to close a 5.5-mile gap remaining in the fence. Both Whitesinger and Smith were among those who successfully stopped the crew. In the background, the Big Mountain Legal Defense/Offense Committee had formed. The legal committee pressed Senator Barry Goldwater to repeal the federal relocation law, which was labeled "an unprecedented disaster" by Leon Berger, director of the Relocation Commission. In July 1984, Senator Goldwater, one of the original sponsors of the Navajo-Hopi Land Settlement Act, introduced an amendment to a Zuni land bill to halt the fencing of the Big Mountain area for two years, pending negotiations with the Hopi Tribe. The legislative efforts failed, however, promising more trouble. Since 1976 only one quarter of the people living on the affected land have been moved. Those who have remained have not been allowed to maintain fences or sheds or to put up new buildings. Many have had their sheep herds taken away from them. A major confrontation was expected by the removal deadline of July 1986. Government officials said they may use federal marshals and National Guard to remove people who resist. Whitesinger expressed the views of many of the Navajo and Hopi people in saying, "I will never leave this land. If they come to move me, they can shoot me right here."

Much of the energy development in the Four Corners area is already underway, particularly at Black Mesa, thirty miles north of Big Mountain. The area is estimated to contain 21 billion tons of coal. The Peabody Coal Company, infamous for its practices in Appalachia, signed leases with the Hopi and Navajo Tribal Councils in the mid-1960s for about 65,000 acres of land with 2 billion tons of coal in the northern part of the Joint Use Area. But traditional Hopi leaders went to court to challenge the leasing agreements. The suit charged that the Hopi Tribal Council was without jurisdiction, authority, or power to issue licenses and that the land belonged to the Hopi villages. In 1975 the Supreme Court dismissed this suit without a hearing, permitting Peabody Coal to proceed with one of the ugliest ecological disasters of our time.

The semiarid desert of the Four Corners area surrounding Big Mountain and Black Mesa is one of the most environmentally sensitive areas of the United States. The National Academy of Sciences has reported that it is virtually impossible to reclaim such land after strip-mining and has recommended that such areas be declared "national sacrifice" areas.

To many observers the Navajo nation has become an energy colony of major American corporations. The power from the strip-mining and massive power plants helps supply electricity to Los Angeles, Las Vegas, Phoenix, and Tucson. In 1984 there were four huge strip mines in operation, five giant coal-fired plants, thirty-eight uranium mines and six uranium mills in and surrounding the Navajo nation. Ironically, the Native American people of the region continue to live in extreme poverty. An estimated 15 percent of the Navajo households have no electricity. And worse are the dangers they face from the radioactive wastes of uranium mining and processing. In 1977, the Navajo community of Church Rock, New Mexico, was the scene of the largest known spill of radioactive material in the United States. It flooded the Rio Puerco, the sole water source for local Navajo ranchers, with 100 million gallons of radioactive mud, sand, acid, and water. Subsequent studies have shown a very high percentage of birth defects in the area, and residents who work in the uranium mining and processing industry are suffering high cancer rates.

To the north in the Black Hills of South Dakota, where two dozen energy companies have staked out a million acres for development, tensions have likewise soared. A coalition of Native Americans, ranchers, and environmentalists joined together in 1978 as the Black Hills Alliance to oppose plans of the energy companies to make part of the Black Hills another national sacrifice area. The stall of the growth of the industry, which forced the cancellations of nuclear plants, has slowed the advance of mining in the Black Hills. And in 1984, nuclear opponents and environmentalists successfully passed a statewide referendum to require that nuclear waste storage questions be brought before the voters. Both this issue and the broader controversy over development of the Black Hills are expected to return if pressures for coal and nuclear expansion heighten in the early 1990s.

Similar to the fencing of land for mining, to help feed expansion of the power empire, is the construction of high-voltage power corridors. Wide-ranging disputes over the construction of extra-high-voltage power lines have produced confrontations as volatile as those over strip-mining. While the struggle for control at places like Big Mountain and the Black Hills are being fought out, rural people in more than a dozen states have rebelled against construction of these lines. Studies have found they produce health hazards for people living or working near them.

Strung from steel towers ranging in height from 150 to 180 feet, the lines hum and crackle with electricity like the sound of bacon sizzling. The lines generate electromagnetic fields so strong that a person standing on the ground below can receive a shock from touching a sizable metal surface, such as the side of a truck or tractor. Adverse physical effects of extra-high-voltage lines on workers in the Soviet Union led that country to establish exposure limits to the lines. Parallel studies both in the United States and Europe have yielded data urging similar precautions.

Dr. Andrew Marino, a research biophysicist at Syracuse Veterans Hospital, asserted in 1976 that exposure to electromagnetic fields can cause altered pulse and blood pressures, fatigue, headaches, and malfunction of vital organs and glands. Also affected, Marino said, are the cardiovascular and central nervous systems. Although the industry's Electric Power Research Institute has maintained that electric fields cause no biological effects of significance, the staff of the New York Public Service Commission recommended to the full commission in 1977 that people who work on or near extra-high-voltage lines should be cautioned against chronic exposure.

More recent research done in the United States and Switzerland has produced evidence linking extra-high-voltage lines to cancer. And a study sponsored by Swedish utilities confirms earlier findings that workers continuously exposed to extra-high-voltage lines run a high risk of genetic damage—with figures showing twice as many deformed children in the families of these workers as found in the general population. In addition to this array of effects, landowners near the line face the grim prospect of soil and ground-water contamination from herbicides used to keep power line rights-of-way clear of weeds and underbrush. In some cases, strong disputes have broken out over toxic chemicals, similar to Agent Orange, used to defoliate the power corridors. All of this has added to a growing controversy.

In upstate New York, when plans emerged for the Power Authority of the State of New York (PASNY) to build a 150-mile extra-high-voltage power corridor in 1973, diverse groups of farming families, Mohawks, land use planners, students, ministers, professors, and small businessmen began forming loose coalitions to stop construction. The proposed 765,000-volt alternating current line would carry electricity generated in Canada to Fort Covington, New York, then south to Marcy, near Utica, New York. The line was capable of carrying 4,000 megawatts of electricity, five times as much as PASNY had contracted for from Quebec Hydro; the remaining capacity was intended to carry power from "nuclear power parks" that were on the drawing boards for the sparsely populated North Country region of New York.

Local citizen groups that formed to stop the line took on names like Upstate People for Safe Energy and Technology (UPSET) and Stop Harmful Occurrences of Cable Kilowatts (SHOCK) and at first pursued their cause through appropriate legal channels, appealing to the courts for relief. When they realized that such well-intentioned efforts were caught in larger politics involving the plans, they turned to nonviolent direct action. Around the time of the first Seabrook demonstrations, people began strapping themselves to tractors and power poles, lying down in the path of construction equipment, and placing farm equipment in the way of utility workers in order to block construction. The growing confrontation reached its climax in January 1977 at the foot of a stately ninety-five-year-old elm tree on the Barse family's farm in Fort Covington.

For five days Stella Barse, who was sixty-five years old at the time and nearly blind, and her neighbors fended off the Power Authority's efforts to destroy this giant symbol of the farmers' resistance. State troopers and power crews, working together, used a cherry picker to serve injunctions and pluck protestors from the tree's upper branches. By the time the heated clash was over, fifteen people had been arrested. Their incarceration sparked a 1,000-person protest march along the proposed path of the line in the vicinity of Edwards, New York.

The major portion of the line was completed and activated in September 1978. In the wake of the dispute, the chairman of PASNY agreed to scuttle all plans for more extra-high-voltage lines during his tenure. Citizen groups continued to stall extensions of the line in 1985, and some rerouting had been won, but more new lines were on the planning horizon.

About the same time the power-line demonstrations erupted in upstate New York, farmers and environmentalists in western Minnesota joined together in equally dramatic opposition to a 435-mile 400,000-volt direct current line planned to run from a large generating plant at a lignite coal strip mine near Bismarck, North Dakota, to the outskirts of Minneapolis–St. Paul. Direct current lines produce a stronger charge than alternating current lines. Fearing controversial health effects, they quietly began to test legal arguments against the line in 1974. Legal opposition provided little success. On June 8, 1976, five days after a construction permit for the project was approved, Virgil Fuchs of Elrosa, Minnesota, took matters into his own hands.

As surveyors were working their way across Fuch's land early in the morning, he drove his tractors toward them and smashed a tripod. He then rammed one of the company's pickup trucks. Fuchs was subsequently summoned to the courthouse, arrested, and charged with two felony counts carrying a maximum penalty of five years and a $5,000 fine. His arrest united protestors and pointed the Minnesota citizen opposition in a new direction, for the first time bringing together farmers from different counties who opposed the line.

When surveyors arrived the following week, more than fifty farmers, notified by CB radio, turned out within a short time to block the surveyors' work. The surveyors left as the farmers began to gather. The Stearns County sheriff refused to take sides in the dispute.

Similar anger surfaced in Pope County a few months later, when survey crews came onto Scott Jenks's land near Lowry. On November 5, about fifty farmers confronted surveyors at a spot on the farm they named Constitution Hill. They set up a picket line with their trucks, blocking the surveyors' path. When the sheriff threatened to arrest one farmer, the farmer drove off, and another pulled up and took his place. After several days of being barricaded and having access roads dug up and cut off, the surveying crew arrived with the sheriff and a court order. Farmers refused to listen as the sheriff read the court order through a bullhorn. Instead they marched into the field singing "America the Beautiful" and the "Star-Spangled Banner."

In early 1978 the situation built to a crescendo when 215 state troopers were used to keep angry farmers from chasing survey and construction crews from a farmer's field. On January 9, in sub-zero weather, two hundred protestors marched to a tower-assembling yard west of Lowry to confront the power-line crew as they had before. Media from all over the region were gathered at the site. When the state troopers blocked the path of the marchers, they stopped and gave the troopers homemade cookies and plastic flowers. There was no violence, but efforts to pass the police barricade were unsuccessful. Leaders of the group said a clash with the state police made no sense and that as long as the power companies had the backing of the state, they would be difficult to stop. Others in the group disagreed. It was a disheartening defeat, but it was not the end of opposition.

As construction of the line continued, various forms of sabotage emerged. Nocturnal "bolt weevils," as they became known locally, took apart bracing for the sections of the 150-foot-high transmission towers and toppled them. In the space of two years, fifteen towers were torn down. Gunfire, assumed to be from hunting rifles, shattered nearly 10,000 glass insulators as the line became a target for local frustration. As in New York, an extension of the line, a sixty-five-mile 345,000-volt alternating current corridor from Delano to Mankato, was canceled in 1983.

One of the deepest ironies in the conflict was that the line was being built by two Minnesota rural electric cooperative organizations, the United Power Association and the Cooperative Power Association. Both are federations of thirty-four small rural cooperatives and thus are technically owned in part by the farmers who were trying to stop the line. Like their counterparts in New York, who had no control over the Power Authority of the State of New York, the farmers had no voice in the decision to build the lines. The policies of public and rural cooperative systems, which had been forced to tie into the power grids by economic reality, were dominated by private companies controlling those grids. A build-or-die mentality was universal.

Generation and transmission cooperatives like United Power and Cooperative Power Association became big businesses run by professional managers who had little sense of the social mission and democratic process that was traditionally a key part of the rural cooperatives. Paradoxically, in their zest to take part in such large centralized power schemes and keep rates down, the managers of the generation and transmission cooperatives often led the co-ops into policies or projects that caused rates to skyrocket. In Minnesota, the electric bills of the 300,000 rural cooperative households doubled when the line began operation in August 1979.

Critics of the lines argued that the destruction of valuable farmland and the health hazards from the extra-high-voltage transmission are often unnecessary and could be avoided if power was generated closer to demand centers, or if conservation and renewable energy strategies were considered as an alternative to such projects.

Battles over extra-high-voltage lines have been fought not only in western Minnesota and New York, but also in Colorado, Florida, Ohio, Pennsylvania, South Dakota, Texas, and Virginia. One of the latest fights in the widespread rebellion against the lines surfaced over mammoth lines proposed for New England. Vermont Governor Richard Snelling agreed to a plan in 1983 to have a 450,000-volt direct current line, the most powerful in the country, transmit power from La Grande River hydro project in Quebec through Vermont's Northeast Kingdom to Barnet, Vermont. The line will eventually run through Monroe, New Hampshire, to Groton, Massachusetts, where it will be converted from direct current to alternating current and transmitted over a 345,000-volt line to a Boston Edison station in Medway, Massachusetts. Members of the Powerline Awareness Campaign—a loose coalition of New Hampshire farmers, environmentalists, housewives, and professionals—have staged opposition to the line, testifying before regulatory proceedings and appealing to local town meetings to pass resolutions against the line.

More than 22,000 additional miles of line above 230 kilovolts are planned for construction between 1985 and 1994. Across the nation's northern border, as many as seven major lines from Canada may be in the offing, leading critics to charge that the future of the nation's power industry is being harnessed to a vision of large centralized power systems. The trend for "wheeling" surplus power in the late 1980s and the construction of these new lines could position the industry for a wave of consolidation and mergers as similar line construction and formation of the power pools had done in the early 1960s.

As serious—or perhaps even more serious—as the damage and risks facing communities from such centralized power industry expansion is the threat to civil liberties and democratic processes. Oftentimes those who have attempted to reveal facts or stood in the way of power industry plans, even scientists such as John Gofman, have faced harassment, intimidation, or ongoing surveillance from investigators employed by the industry.

Danny Sheehan, a Washington, D.C., based attorney, claims that these corporate tactics are undermining basic civil liberties and hold far-reaching threats for the future. In a nationally publicized suit, Sheehan represented the family of Karen Silkwood, a Kerr-McGee Corporation employee suspected of having been murdered in November 1974 while on her way to provide a *New York Times* reporter with documentation on violations of safety standards within the nuclear industry. In investigating the Silkwood case, Sheehan's legal team uncovered a pattern of surveillance and espionage used by power companies and the nuclear industry all across the country, which local or state police had condoned or participated in.

"Although we were trying to investigate what went on out near Oklahoma City," Sheehan says, "we were told by an inside source to start looking in Georgia." At the Georgia Power Company they found a surveillance unit being run as part of the company's public relations department. The unit had been started in 1973 and was modeled after the security departments of several other utility companies, including Southern California Edison, Pacific Gas and Electric, and the Alabama Power Company. Although security for nuclear plants is essential, the scope of Georgia Power's activities, as well as those of other private companies, went well beyond ensuring safety at nuclear plant sites.

Two former security investigators for Georgia Power, William Lovin and John Taylor, were interviewed by NBC news reporter Paul Altmeyer in December 1977. They revealed that individuals and organizations who "for any reason would be against the rate increases or would have some type of critical opposition of the power company" were investigated. "The practices were deliberately designed to identify protest groups and individuals . . . if security was ordered to investigate they would investigate, clandestine or otherwise." When asked to explain what the investigations entailed, Lovin declined to answer. "Because one, it's illegal; it is improper," he said.

In a second part of his investigative report, Altmeyer interviewed Jerry Ducote, a former deputy sheriff and a California district leader for the John Birch Society who admitted to seventeen burglaries. Ducote said he worked for a private investigation firm known as Western Research (later changed to Research West) that handled a number of power company contracts. Ducote said that information taken in two break-ins was given to the Pacific Gas and Electric Company. One involved taking a mailing list and other documents from the home of an old woman who was a critic of Pacific Gas and Electric policies. Although the company denied the allegation, Federal Power Commission records showed that from 1971 to 1976 Pacific Gas and Electric had paid $90,000 for "investigative services" to Research West.

With a sense of the national scope of the problem, Sheehan's investigators followed their trail to Florida, where several of Georgia Power's security people had been trained at the National Intelligence Academy, a CIA center near Fort Lauderdale. Sheehan said what was happening was that former special police and federal intelligence people who had been trained at the center were taking jobs with power companies or with private security operations holding contracts with power companies. "They had the cooperation of local police and federal intelligence people because they knew them," Sheehan said. "They could go through the national computers at the Law Enforcement Intelligence Unit, or they could connect into the government's information systems and get anything they needed, just from tracing a license plate or getting a name."

A folder at the Fort Lauderdale center showed that members of a special police unit for Oklahoma City had trained there. After locating Oklahoma City's special police unit, investigators found that its former director, James Reading, was head of Kerr-McGee security. Sheehan said evidence turned up that Kerr-McGee, the local special police unit, the state police, and the FBI conducted wiretaps and surveillance of Silkwood before her death. Although there was insufficient evidence for a murder charge, Sheehan sued Kerr-McGee for violating Karen Silkwood's civil rights and the FBI for covering up the conspiracy. The courts refused to hear the case, ruling that civil rights laws covered only discrimination against ethnic or racial minorities.

As evidence showing wiretaps and surveillance placed on Karen Silkwood emerged, attention began to focus on the problem. Other reports exposed power company surveillance and harassment of critics in at least eighteen states. The revelations raised a storm of concern, including a 1979 congressional investigation and a flurry of media reports. In New Jersey, where Jersey Central Power was spending $500,000 a year on "security," the conflict was over photographing of license plates and individuals. In Seabrook, New Hampshire, and on the opposite coast in Los Angeles, California, it was wiretaps, surveillance, and videotaping of activists and opponents to power company plans. In Philadelphia it was the keeping of files and dossiers on antinuclear activists by the Philadelphia Electric Company. In New York it was infiltration of citizen groups and a $2 million suit filed by Long Island Lighting Company against forty-one individuals. In California, a similar $1 million suit was filed against opponents of the Diablo Canyon nuclear plant. At a symposium sponsored by New York University in September 1981, Jay Peterzel of the Center for National Security Studies said that from 1971 to 1981 there were ninety-three recorded incidents of break-ins, wiretaps, surveillance, and harassment of power company critics. Because few such incidents are formally reported, the actual number is assumed to be much higher.

Technically, there is no law against private companies conducting surveillance or keeping files on individuals. The legal system has yet to recognize any significant limits on private company activities to trample on civil liberties. Danny Sheehan says that the loophole presents a huge problem. The power companies and local police justify their actions as being in the interests of economic and national security, but they are blurring the distinctions between the public's need for power and the protection of specific private companies and their plans for centralized growth. "This is 'Big Brother,' " Sheehan says; "these companies can't be allowed to harass people for political reasons. If people didn't have the right to protest about nuclear power plants, we'd have hundreds of them operating now instead of eighty or so. And you can bet there would have been accidents worse than Three Mile Island."

Until the Chernobyl disaster in the Soviet Union in April 1986, Three Mile Island was the case in which the danger facing communities received its most visible recognition. Although the Soviet accident was much more severe, there were many disturbing parallels over uncertainty and delay in informing the public of risks.

□ □ □

For the people living near Harrisburg, Pennsylvania, March 30, 1979, was Black Friday. It marked a day filled with frightening announcements concerning Metropolitan Edison's nuclear reactor perched on Three Mile Island in the middle of the Susquehanna River. That morning, two days after the start of the worst commercial nuclear accident in America's history, bursts of radioactive gases were released from the crippled Unit 2 and blown by the wind toward neighboring towns. Area residents received warnings to stay indoors. Two hours later Pennsylvania Governor Richard Thornburgh appeared in a nationally televised press conference and advised pregnant mothers and families with preschool children living in the vicinity to evacuate. Behind the scenes officials debated whether or not to order wider evacuations.

Release of radioactive gases posed a serious health threat, but the most alarming news was the discovery that a potentially explosive hydrogen bubble had formed in the reactor—a development that conjured up the specter of a core meltdown and large-scale release of radioactive material. Although both state and federal officials tried to play down the seriousness of the accident, public suspicion spread (and government transcripts later confirmed) that the experts were not in control of the situation. That Black Friday morning, more than forty-eight hours into the crisis, NRC Chairman Joseph Hendrie told his weary staff, "We are operating almost totally in the blind, his [Governor Thornburgh's] information is ambiguous, mine is nonexistent . . . it's like a couple of blind men staggering around making decisions." Over the weekend, an estimated 200,000 residents fled the area, many in response to Governor Thornburgh's cautionary advice and others out of a fear that the situation would worsen.

The accident had begun in the early morning hours of Wednesday, March 28, 1979, when a roar of steam went up into the 4 A.M. silence surrounding the plant. Because of a problem with a safety valve connected to feed-water pumps, the cooling system had malfunctioned and the power turbines shut down. Routine emergency procedures were automatically initiated, and within seconds, three emergency feed-water pumps went into operation. The incident should have ended there, but someone had closed two of the valves through which the feed-water was supposed to flow. A chain of human and mechanical failures followed, including a stuck relief valve that drained the coolant water below safe levels. For more than sixteen hours, the hot uranium fuel core was inadequately cooled. Thousands of gallons of cooling water drained from the reactor core, and the water inside began to flash into steam, blanketing much of the fuel rods. The net result was a partial core meltdown. As the fuel core overheated, swelled, and ruptured, large amounts of radioactive material were released into the containment building.

The first news of the accident was met with disbelief. The following day, Thursday, radio and television news reporters who converged on the site maintained coverage, presenting conflicting reports from state, company, and federal officials. Throughout eastern Pennsylvania, New Jersey, New York, New England, and as far south as Washington, D.C., people watched and listened to media reports, fearing that if a large release of radiation occurred, winds could carry a radioactive plume hundreds of miles to the east or southeast, a possibility studies dating back to the 1950s had outlined. In many towns people, listening attentively to the news broadcasts, made preparations to evacuate should a sudden need arise. For five days the atmosphere was one of chaos.

It wasn't until Sunday night that the NRC staff discovered that fear of a hydrogen explosion was based on spurious technical analysis. Information now showed that the "bubble" didn't exist. The tension eased, but it was not until nearly a month later, on April 27, that the plant was finally stabilized and put into controlled shutdown conditions. Radioactive gases and nearly a million gallons of radioactive water remained inside the containment building. The cleanup was at first estimated at $430 million and was expected to take four years to complete. Seven years later, in 1986, cleanup was still continuing with a price tag exceeding $1 billion, much of which was being subsidized by the federal government.

For the people living in the area surrounding Three Mile Island, the plant remained a symbol of horror and tragedy, far from the "bright and shining option" of a nuclear-powered future envisioned at its dedication ceremony in September 1978. In a larger arena, the accident marked a major shift in the struggle over nuclear power. It shattered the myth of the "peaceful atom" and proved that despite the astronomical odds nuclear supporters had placed against an accident happening, such occurences were a very real possibility.

Ironically, the Three Mile Island accident came at a time of heightened public concern over reactor safety. Just two months before, in January 1979, the NRC had withdrawn its endorsement of the 1974 Rasmussen report, a document the industry relied upon to quell public fears about nuclear accidents. Like the authors of the *WASH-740* reports, Dr. Norman Rasmussen, an ardent nuclear supporter, rated the chances of a serious accident at about the same probability as a meteor hitting a major city (one in a million). A few weeks after withdrawing endorsement of the Rasmussen report, the commission shook the industry's confidence more by ordering five operating nuclear plants shut down and modified because of design miscalculations, which made them susceptible to damage from earthquakes.

Even more ironic than the NRC actions was the release of the film *The China Syndrome* just weeks before the accident. This fictional account of a possible nuclear meltdown bore an eerie resemblance to what actually happened at Three Mile Island. The reassuring statements issued by company officials sounded as if they were taken right out of the movie script. Jack Herbein, Metropolitan Edison's vice-president for power generation, for example, dismissed the incident at first as merely "a normal aberration." Other company officials reassured the public that there was no danger of a radioactive release—even as plans were being made to take the risk of venting radioactive gas into the atmosphere.

Although a major catastrophe was narrowly avoided at Three Mile Island, the industry suffered a credibility loss of immense proportions. Millions of Americans who had never before given much thought to nuclear power received a crash course on the subject. Polls showed that 98 percent of all Americans had heard about the accident and that most didn't believe they were being told the whole story. A joint CBS/*New York Times* survey revealed even more ominous news: Only 46 percent of the people now favored further development of nuclear power, compared to 69 percent who were asked the same question in July 1977, a few months after the massive Seabrook demonstration.

Public response to the accident put new life in the antinuclear movement, stirring thousands of people from diverse backgrounds to oppose construction and operation of nuclear plants. New initiatives proposing licensing and construction moratoriums were introduced in a dozen states, and intervenors in regulatory proceedings added emergency safety and economic responsibility issues to their motions. In May, the National Council of Churches, the nation's largest ecumenical organization, called for a new national energy policy that would shift dependence on nuclear power toward conservation and renewable energy resources. By a 120–26 vote, the governing board of the church group supported a de-emphasis on nuclear development, because of unresolved waste problems, the possibility of catastrophic nuclear accidents that would permanently contaminate large areas of land, and possible "irreversible damage to the environment and to the human genetic pool."

The accident also drew attention to scores of demonstrations and sparked others. On March 31, the third day of the accident, 100,000 people from all over Europe came to northern Germany to protest the West German government's plan to build a huge radioactive waste storage and reprocessing plant in the small farming village of Gorleben. Two weeks later, 1,500 people attended a rally in Harrisburg, which focused on nuclear safety issues and who should pay for the cleanup at Three Mile Island. The following weekend, 15,000 people turned out for a protest rally at Rocky Flats, a nuclear-weapons plant outside of Denver, while a coalition of Native American tribal councils, Chicano community organizations, and antinuclear alliances staged three days of protest against uranium mining on Native American lands in New Mexico. A national march and rally turned out 100,000 protestors against nuclear power and nuclear weapons on the steps of the Capitol in Washington, D.C., on May 6, 1979.

In the fall, more than 300,000 people participated in a week-long series of events in New York City, which was highlighted by five nights of music in Madison Square Garden, featuring James Taylor, Bonnie Raitt, Bruce Springsteen, Jackson Browne, Gil Scott Heron, John Hall, and other performers who donated their services as Musicians United for Safe Energy to raise funds for antinuclear groups. A month later, in October, 10,000 protestors from New England and mid-Atlantic states participated in a Wall Street protest to mark the fiftieth anniversary of the 1929 stock market crash and to expose the role of the financial community in bankrolling the highly capital-intensive development of nuclear power. It was an attempt to focus attention on the re-emergence of the new power empire. Mapped out by the New Hampshire Clamshell Alliance and New York City's Shad Alliance, the Take It to Wall Street demonstration was endorsed by forty-four community and activist organizations. Along with a rally, the protest included nonviolent civil disobedience to block the entrances to the New York Stock Exchange on Monday, October 29.

Although the new power empire suffered a major setback in the shift of public sentiment against construction of reactors, it was able through quick, decisive, and well-funded action to limit the political fallout from Three Mile Island. With its tremendous influence in Congress, the power industry and its corporate allies defeated every major piece of legislation challenging the future of nuclear power. None of the ten bills introduced in Congress in response to the accident, calling for a moratorium for either a fixed period or until the safety of nuclear power could be guaranteed, ever made it to the floor of either chamber. Not until Representative Edward Markey of Massachusetts introduced his moratorium proposal as a floor amendment in November 1979 was the whole House finally dragged to a vote on the issue. As a clear measure of its strength, the industry marshaled enough support to crush the amendment by a vote of 254 to 135.

Using its age-old press strategies, the power empire also attempted to misrepresent key findings and observations on the accident. The industry claimed that the Kemeny Commission, put together by President Carter to investigate the accident, found that nuclear power development should "proceed, but with caution." In actuality the commission took no such position, but stated in pallid bureaucratese that "if the country wishes, for larger reasons, to confront the risks that are inherently associated with nuclear power, fundamental changes are necessary if those risks are to be kept within tolerable limits." Given the deficiencies uncovered in reactor design, shortcomings in regulatory oversight, inadequate operator training, and serious problems in other areas of day-to-day operation, the commission said, "an accident like Three Mile Island was eventually inevitable." Contrary to the message interpreted by the industry's propagandists, the commission noted that, in order to prevent accidents as severe as Three Mile Island in the future, "fundamental changes will be necessary in the organization, procedures and practices—and above all—in the attitudes of the Nuclear Regulatory Commission and, to the extent that the institutions we investigated are typical, of the nuclear industry."

Despite these criticisms virtually nothing was done to carry out the recommendations made for fundamental changes in regulating the nuclear industry. Operator training was upgraded, and new safety procedures were put in place, but the change in attitude and the suggested restructure of the NRC and efforts to put someone in charge "unwedded" to the industry were not carried out. Nunzio Palladino, a Westinghouse engineer and former member of Governor Thornburgh's energy council, took over as head of the NRC in 1982. He provided insight into what the industry strategy would be in the 1980s. In a speech given in 1979 at a conference on radiation and health sponsored by Pennsylvania State University, Palladino prescribed more propaganda. He said, "Public and political overreaction [to the Three Mile Island accident] will be detrimental to nuclear power, and a big job of public education must be faced if nuclear power is to be part of our future."

Nuclear manufacturers and power companies had already embarked on a major advertising campaign. Over a six-month period following the Three Mile Island accident, the industry spent $1.6 million to bolster the sagging image of the "peaceful atom." This was only the beginning. The planning, guidance, and coordination for a larger "public education campaign" would be entrusted to the U.S. Committee for Energy Awareness, an organization launched at a meeting held at Bechtel headquarters in May 1979. Members of the propaganda committee would include GE, Westinghouse, the Atomic Industrial Forum, Bechtel, and Combustion Engineering, and the major electric power companies. The EEI would take on the role of overseeing the development of the committee and collecting its funding from utilities as the National Electric Light Association had done in the propaganda campaigns of the 1920s.

A huge media battle was shaping up. Representatives of the power empire called it a fight for the "hearts and minds" of the American people. Slick multi-million-dollar advertising campaigns coupled with polling, lobbying, and public speakers were aimed at creating a shift in public attitude to support pronuclear legislation and prepare for a "second coming" of nuclear construction in the late 1980s or early 1990s. Among some of the many similarities to the industry propaganda campaigns of the 1920s was a program to "educate" school children about nuclear energy.

In the summer of 1983, Congressman Richard Ottinger of New York held a House subcommittee hearing to investigate contributions and activities of the Committee for Energy Awareness (CEA). Part of the investigation revealed that the educational program had provided $600,000 to a group known as Scientists and Engineers for Secure Energy during 1979–83 for pronuclear seminars at colleges and other public speaking engagements. Through surveys and polling, the industry had found that the public was more inclined to believe scientists than power company officials or antinuclear activists. Mark Mills, a physicist who opened the Washington, D.C., office of Scientists and Engineers for Secure Energy, had previously been employed by the Atomic Industrial Forum. Ottinger also discovered that taxpayers had become unknowing contributors to the pronuclear scientists' organization. The Department of Energy had granted $100,000 to the scientists' group in 1982 for pronuclear college forums. Other money was also quietly channeled to the propaganda campaign.

Funds slated for a broad magazine and television advertising campaign scheduled by the Committee for Energy Awareness totaled $24 million in 1983 and $24 million in 1984. Congressmen found that many power companies contributing to the campaign were passing the costs of their propaganda through to consumers. Everard Munsey of the Virginia Electric Power Company defended this practice and said that if Congress sought to limit this charge, it would limit the power companies "in the exercise of their first amendment rights." Outraged by such a claim, subcommittee member Ron Wyden of Oregon replied, "That's ridiculous."

The subcommittee pressed Harold Finger, director of CEA and a former General Electric executive, to release a list of the power companies making contributions and the amounts donated. A partial list was released showing Westinghouse and GE each donating over $1 million, Commonwealth Edison of Chicago donating $3.5 million, Pacific Gas and Electric contributing $789,000 in 1983, Public Service Company of New Hampshire donating $266,650 in 1983, and as many as ninety other power industry companies also supporting the program.

The irony of this propaganda campaign was not lost on the people of Pennsylvania, who watched the advertising about "safe nuclear power" on national television networks. Governor Richard Thornburgh criticized the power industry in the fall of 1983 for paying for expensive advertising while ignoring the urgent business of cleanup of the accident at Three Mile Island. "Every day that goes by increases the delay," Thornburgh said, "and is a threat to the health and safety of the populace." The power industry had promised $150 million toward the $1 billion cleanup effort. An EEI spokesperson said that the institute was "disappointed" in the reluctance of members to contribute to the cleanup fund, which the federal government (taxpayers) was being forced to subsidize.

In the early 1980s the industry was spending more than $20 million per year in a national advertising program to convince the public to support continued power industry expansion. The industry targeted heavy advertising for cities that faced controversies over rate increases from nearly completed nuclear plants and also ran advertising on national television networks. The primary advertisement run by the Committee for Energy Awareness in 1983 and 1984 featured the theme from the popular Broadway hit *Annie.* Amid a montage of sunsets comes a little girl's voice singing, "Tomorrow, tomorrow . . ." as a narrator explains that nuclear power is the only reliable energy source and other alternatives are way off in the future. Another ad later developed and sponsored by the CEA's lobbying sister, EEI, highlighted the importance of electricity in modern life and alluded to a "freedom of choice" consumers have in using or conserving electricity. But the "freedom" was a mirage. The real choices over what sources that power would come from and the extent of conservation to be undertaken were left to the industry. The advertising featured in 1986 stressed the need for "energy independence" that could be provided by nuclear plants. The irony was that a significant portion of the uranium used for fuel in U.S. nuclear plants was imported, and with only 4 percent of electricity generated by oil-fired plants in 1985, nuclear plants displaced very little oil imported from foreign countries.

Antinuclear activists and consumer groups countered the power empire's radio and television propaganda by fighting for equal time for renewable energy advertising. By submitting requests to television and radio stations for equal time under the Federal Communication Commission's fairness doctrine, local grass-roots groups in Phoenix, St. Louis, Chicago, Portland, Maine, Greenville, South Carolina, Rochester, New York, and a half dozen other major cities presented radio or television advertising to counter the power industry "blitzes." Scott Denman of the Safe Energy Communication Council, a public interest organization based in Washington, D.C., which helped local citizens counter the CEA campaign, said the industry's advertising has been effective in dampening the opposition to nuclear power and in undermining support for solar as well.

□ □ □

By the early 1980s the handwriting was on the wall. The drive begun in the 1960s to triple the output of power that had brought about a reemergence of the power empire also had created an enormous financial disaster. Between 1974 and 1986, 103 nuclear reactors had been canceled. No new orders had been placed since 1978. The enormous influence wielded over federal and state regulation had backfired. Power demand that the industry had projected to grow at the traditional rate of 7 percent per year had slowed to an average of 2 percent. The plants the power industry had fought so fiercely for would now only add to a burgeoning power surplus that reached nearly 40 percent in 1982. Overbuilding, overfinancing, and soaring costs forced the power industry into a stall as it clung to its vision of a highly centralized future.

Angered by the sacrifice of a safe environment, the sacrifice of civil rights, and the sacrifice of economic security to the power empire, and thoroughly disillusioned with politicians who still supported the industry, consumers turned to the ballot box to reassert public control. In 1982 citizen groups in eight states used the referendum process to place questions regarding power companies before voters. The success was mixed. In Maine, a proposal to shut down the Maine Yankee nuclear plant within five years of the vote drew strong opposition from the power industry. More than $800,000 was spent in an advertising blitz by power industry allies to defeat the measure. But in Massachusetts, voters gave overwhelming support to a ban on construction of any new nuclear plant or radioactive waste disposal site, unless the proposed project meets vigorous environmental, safety, and economic conditions and gains the approval of voters in a statewide referendum. Organizers found that, even if the issue was sound, it took a very careful gathering of support and a significant amount of capital to win. Back-room politics often became the deciding factor. They also discovered that the power empire had developed a formula approach to ballot box campaigns.

Winner-Wagner, a New York public relations firm, specializes in spearheading the power industry's campaigns. The firm boasts a 7–1 record in defeating major citizen initiatives. Carrying the methods first developed by industry public relations men like Ivy Lee into the computer age, the industry's propagandists make an art of manipulating public opinion. Through carefully orchestrated advertising, the use of power company front groups made up of bankers, power company personnel, and other business leaders, and sophisticated tracking polls, Winner-Wagner has been able to turn popular opinion around even when the odds stand at two or three to one against their clients.

In Missouri in 1984, for example, a broad coalition of citizen groups and state legislators proposed an Electric Ratepayer Act, which sought to provide state regulatory officials with the authority to require that power companies show that any power they deliver is the cheapest available. It would have prevented future overbuilding and required thorough conservation programs to be undertaken by the state's power companies. Faced with rate increases of 60 percent from the Calloway and Wolf Creek nuclear plants, consumers readily signed petitions to have the question placed on the ballot. What happened from there is a different story.

More than twenty-eight power companies from eighteen states contributed to a war chest of $3 million to help the Missouri companies defeat the proposal. With this money and the aid of Winner-Wagner, the power companies saturated the air waves with messages that the proposal would not lower but raise electric rates. Each household, the ads said, would face an increase of $1,800 per year, the result of the Calloway nuclear plant (whose power is not needed) being abandoned if the question passed. Power company attorneys also filed suit against the proposal, keeping its status in limbo until just thirteen days before the election. The final result was a 2–1 vote against the act.

Other initiatives in West Virginia, Iowa, Michigan, and Oregon met with greater success, showing that it is possible to win referenda with strong grass-roots participation, but stiff industry resistance can be expected. Another ballot box trend has been in the fiery efforts to regain citizen control of public power and rural cooperative systems. The dramatic Washington Public Power Supply System ratepayer rebellion in the Northwest is the most widely known example of this. But all across the country from Austin, Texas, to Montpelier, Vermont, the same issue has surfaced, spurred by soaring costs from investments in nuclear and coal projects. In those public power areas, the hope for citizen control over an alternative future for electric power supply has stirred the democratic process and helped bring about surprising results.

In 1979, thirty-one municipally owned power systems in Massachusetts were offered 13.8 percent joint ownership in the failing Seabrook project. Although municipal power system managers disregarded criticism of the project and were eager to sign contracts, consumer advocates organized statewide opposition. Twelve of the towns held special town meetings and hearings to debate the issue. Ten of them turned it down, and another five reduced the shares they were offered, cutting the purchased amount to 6 percent, ultimately forcing unit two to be abandoned. The summer-long series of meetings were particularly intense. In the town of Groton, Massachusetts, emotions at the town light commissioners' office reached fever pitch over the proposed $3 million purchase. The overflow crowd moved the meeting to a nearby fire station. Part of their anger resulted from the fact that the multi-million-dollar deal had been proceeding behind closed doors with no notice to the public of what was pending. A special town meeting turned down the investment, but the commissioners insisted that the town would go ahead with the purchase, regardless of what citizens wanted. Connie Whitney, a gray-haired woman in her mid-sixties, said she had never been as close to a riot or mob scene as in the firehouse debates. After the meeting, Whitney stopped two of her matronly friends from throwing rocks through the plate glass window of the light department directly across the street. Within a few years, Whitney and other consumer advocates in five towns across the state would be elected as municipal light commissioners, holding promise for long-term changes.

In Hugo, Oklahoma, an equally turbulent meeting took place in November 1982 when nine incumbent directors of the Choctaw Electric Cooperative were ousted from office by angry consumers. Their removal was the culmination of a six-month recall campaign resulting from dramatic increases in rates and allegations of mismanagement. It was the beginning of a wave of identical recalls that swept through the rural co-ops in what was dubbed the Choctaw Syndrome. In April 1983, more than one hundred black members of the Mississippi Delta Electric Power

Association, who felt they weren't getting fair representation, decided to show up at the co-op's annual meeting to do something about it. The incumbent board of directors, all white males in a region that is 75 percent black, panicked, claimed there wasn't a quorum present, and quickly adjourned the meeting while black members were still registering to vote. The insurgent members continued the meeting, found there was a quorum, and proceeded to elect a new eleven-member board. This was the first time since the co-op's inception in 1937 that blacks had held positions on the board. The incumbents who fled the meeting allegedly had not bothered to muster a quorum for an annual meeting since 1967 and simply re-elected themselves or appointed their friends without holding elections. The new commissioners immediately began to examine the rate structure of the co-op, which was said to discriminate against low-income people. They also began to explore the need for conservation and alternative power supply.

A resurgence of citizen control over public power and rural cooperative systems holds the promise of reaffirming the vital social role electricity plays and renewing the original social mission and sense of democracy of these public nonprofit agencies. It also promises to stimulate conservation and decentralized alternative energy development and restore the municipal and cooperative systems as an opposing force and yardstick to the build-and-grow centralization of the private power companies. Many alternative energy advocates point to public power systems in Osage, Iowa; Davis, California; Austin, Texas; or Seattle, Washington, as models for the use of practical solar applications and conservation programs. But given that the public power and cooperative systems are linked into power pools dominated by private power companies, their long-term success in moving away from more centralized technology may remain largely unrealized.

Most people are unaware of the century-long turbulence over the power industry and how the current citizen rebellions fit into that continuum. The underlying issue remains one of control. The popular hope is that solar and renewable energy technology will provide the opportunity to break down monopoly control and give birth to a new competitive, decentralized electric power industry over which consumers will have greater voice. This hope reverses the role of technology, however, by ignoring the fact that it has been politics that has shaped the market conditions and choices in technology throughout the course of the industry. In the view of optimistic alternative energy advocates, an "energy transition" from large centralized power systems dependent on fossil fuel and nuclear power to decentralized systems, using an array of cogeneration, conservation, and renewable energy sources, is already in process and will continue to evolve over the next twenty to fifty years. The example commonly pointed out is Southern California Edison and a general trend toward wind and solar energy development in California. The development in California has brought rapid change. From 1982 to 1986 independent power production increased from 100 megawatts to 2,200 megawatts. As much as 25 percent of the state's electricity is expected to be provided by independent producers by the turn of the century. A few other states are projecting similar growth. Maine, for example, anticipates as much as 30 percent of its power will come from independent producers, largely cogeneration systems at paper mills and small hydro projects. Former Solar Lobby Director Dick Munson claims, "The age of centralized power plants is over." What Munson foresees is "the rise of independent power producers and the rise of competition" as "the real step in the right direction for consumers and for the electric utility industry itself . . ."

But very serious obstacles lie in the way of fulfilling these hopes. Removal of solar, wind, and other tax credits, for example, will undoubtedly slow alternative energy production, and power company opposition to the development of cogeneration has brought myriad problems. In California, a moratorium placed on cogeneration to avoid disruption of power system rates and distribution signals potentially significant delay. The Federal Energy Regulatory Commission has eliminated third party financing of cogeneration projects. And private power company influence over federal and state regulation, and control of major power pools and transmission and distribution systems, could stall development further. In the meantime the industry is rolling out its advertising and trying to gather support for a "second coming" of nuclear power in the early 1990s.

Most executives of private power companies still hold the belief that the prosperity of the nation is dependent on increasing consumption of electricity—on production rather than efficiency. The common industry view is that the energy "transition" will be one simply from oil to coal and nuclear power—a switch in fuels which they promote as creating United States "energy independence." According to the view of history promoted by the EEI, the Industrial Revolution was followed by an Electrical Revolution. In the next wave of this revolution industry executives foresee increasing industrial electrification and automation in the steel industry, the auto industry, textiles, plastics, and dozens of other manufacturing processes. At the community level they foresee more electricity-intensive, climate-controlled facilities such as huge shopping malls, all-electric homes, electric cars, mass rail transportation powered by electricity, and eventually all-electric cities. All of this would be fed by networks of nuclear plants and coal plants and ultimately breeder and fusion reactors and possibly by alternatives that fit with centralized systems, such as giant windmill farms, huge centralized solar facilities, and solar-powered satellite systems. These satellite systems would link millions of solar cells in space stations half as large as Manhattan Island to beam down microwaves to central earth-based receiving stations, from which electricity would be distributed. Feeding this vision in the short-range future means building up the nation's electrical transmission grid and construction of new coal and nuclear plants.

Nuclear Regulatory Commissioner Lando Zech, Jr., told an Edison Electric Institute gathering at Marco Island, Florida, in January 1986, "If the electrical growth in the United States is between two and three percent annually, nearly two hundred new 1,000-megawatt plants will be needed by the year 2000. This means at least one new large power plant, either coal or nuclear, will be needed in our country each month for the next fifteen years."

Critics who argue for productivity rather than production say these inefficient systems will not only do irreparable damage to the environment, centralize the economy, and undermine civil liberties, but will also produce a capital-intensive energy system that will bankrupt the nation. Some believe that the fulfillment of this new wave of electrification is impossible and that market forces simply won't support it. Amory Lovins is among the optimistic alternative energy leaders who believe that a "transition" to solar and conservation strategies is already happening at a very rapid rate. Lovins points to more than a million buildings that have been outfitted with passive solar equipment in the past decade. He also points to a "revolution" in conservation and energy efficiency that promises to bring more small independent power producers and more jobs into an "energy services" market. "What the power companies are competing with in the market place are not solar photovoltaics or more exotic technology but weather-stripping and calking and consumers deciding to insulate homes and use more energy-efficient appliances," he says. "Smart power company managers realize they're not in the business of selling electricity anymore, but in the business of selling 'energy services.' "

All of the electricity growth since 1974 has been met by efficiency improvements and conservation, Lovins says, and that's only the beginning of a much larger supply of energy that could be saved without a change in people's lifestyles. "Those savings opportunities have already made every thermal power plant in the country uncompetitive," he says. "I think that although there is a kind of last hurrah mentality occasionally raising its head, the nuclear option no longer exists."

Scott Sklar, a friend of Lovins and another former director of the Solar Lobby in Washington, D.C., disagrees with Lovins' analysis of how a "transition" is taking place. Sklar says, "Things aren't going to happen just because we want them to happen, or because they make sense. They'll happen because of forces of the marketplace and public opinion. The industry controls those forces right now. They plan on building more central plants, and that's the way everything is headed. . . . Amory says that nuclear is dead. Nuclear is not dead. You're talking about an investment of more than $150 billion. The banks and people behind that are not going to just get up and walk away. There is a group of executives of ten or so utilities that see the importance of solar and conservation, but there is another group, a much larger one, that sees building more centralized nuclear and coal plants as essential to the nation's economy." The fundamental problem, Sklar says, is a political rather than a technological one.

Sklar's views are supported by statements of industry leaders. In a hearing before the House Subcommittee on Energy Conservation and Power, Michehl Gent, head of the North American Electric Reliability Council, told congressmen, "It is our view that the only realistic and dependable options for capacity in the future are new nuclear and coal-fired generating units." Gent said the greatest problem in meeting future power needs was "the long lead time (ten to twelve years) to bring a new generating unit into service from final planning decisions to commercial operation." Instead of new regional planning and regulation being discussed at the hearing, Gent called for a streamlining of the process to license plants to shorten the lead time needed in planning.

The outcome of this debate is still uncertain, and the stall it is causing may have a longer-lasting effect than optimists now realize. According to a major report released in 1985 by the highly respected Congressional Office of Technology Assessment, conventional central stations will probably continue to be the choice of utilities for the 1990s. "Virtually all of the technologies considered in this assessment offer the potential of sizeable deployment in electric power generating application beyond the turn of the century," says the report. But at the current rate of development "most of the technologies will not be in a position to contribute more than a few percent of total U.S. electric generating capacity in the 1990s." In the meantime, as it presses ahead with centralized plans, the industry may be able to stifle the market for decentralized power production further by removing the 50 percent ownership limitations under PURPA, and winning passage of new laws and regulatory changes.

Senator Dan Evans, a member of the Senate Energy Committee and former head of the Northwest Power Council in Washington state, told a joint conference of the Atomic Industrial Forum and the American Nuclear Society in November 1985 that the situation of uncertainty in the utility industry could be described with an eight-letter acronym—TMIWPPSS. The Three Mile Island accident frightened the public and Congress on the safety of nuclear power, and the WPPSS economic problems and default frightened them on costs. Evans vowed to introduce legislation to facilitate regional planning and called for reforms to expedite issuance of permits and licenses for nuclear reactors. He said he hoped the industry would communicate "a sense of urgency" to elected officials.

At the local level, power companies are still encouraging inflated and inefficient use of electricity. For example, more than half of all the new homes being built continue to be electrically heated. While a home owner pays only $1,000 for an electric heating system, Lovins points out that the power company will spend $200,000 on capacity to heat that single home (not to mention the strip-mining or radioactive waste storage or other environmental impacts of generating this power). Lovins says it would make more sense to have the power company supply energy-related services such as helping finance the purchase of energy efficient appliances, lights, weatherization, and solar heating systems, which reduce electrical use and costs and have less environmental impact. But for the new home owner the energy services for $10,000 in solar or other energy alternatives are not in existence. Money must be paid up front, and there are few power company loan programs in place to do it. The political bridge offered by the Public Utility Regulatory Policy Act and other new laws are not being fully utilized.

It is not that the potential is not out there. More than 200,000 megawatts of electricity from cogeneration alone are estimated to be available by the year 2000, according to the Office of Technology Assessment. Already, the Federal Energy Regulatory Commission has industrial sites with 24,533 megawatts of cogeneration potential on file. If the cogeneration potential, coupled with vigorous conservation programs and renewable energy development, is pursued, it appears realistic that growth in electricity demand can be met without constructing new generating plants. However, as Sklar and industry testimony point out, most power companies remain reluctant to pursue these options. Clearly it would cut into the age-old grow-and-build mentality as well as profits and the historical sense of power companies being at the heart of shaping the corporate vision of the nation's all-electric future.

Political control, indeed, remains the determining factor over what kind of an "energy transition" will take place. Believing technology to be the key, alternative energy proponents and consumer advocates have moved the fight for the nation's energy future to state and federal regulatory arenas. Here, they hope to stop the strip-mining and stringing of new extra-high-voltage power lines, and the building of new coal and nuclear plants—and instead ensure vigorous conservation and energy efficiency programs, and institute guidelines for least-cost energy planning. Regulation, however, is thickly wrapped in corporate politics. The state agencies are generally underfunded and understaffed, and the federal agencies are dominated by Reagan appointees who do not believe in regulation. The fundamental question is whether the regulatory agencies possess the will or the capability to break the monopoly control of the industry and select a future different from the centralized vision it has mapped out.

7. Regulation: The Impossible Task

I venture the statement, in true fear of contradiction, that no shrewder piece of political humbuggery and downright fraud has ever been placed upon the statute books. It's supposed to be legislation for the people. In fact, it's legislation for the money oligarchy.

Daniel Hoan, Mayor of Milwaukee, Wisconsin, on the passage of a law for state regulation of private power companies, 1906

From a distance, the long two-story building that houses the New Hampshire Public Utility Commission (NHPUC) could be mistaken for a rundown ten-unit motel off the main highway from Concord. The parking lot used by the agency staff and visitors during the day provides space at night for tenants, who live upstairs on the second floor of the brick structure. Inside, more than fifty-five full-time employees are crammed into cubicles working on a variety of regulatory issues. In addition to overseeing electric and gas companies, the commission also has responsibilities for the state's railroads, bus lines, commercial vehicles, and telephone companies. It's clearly a heavy burden. The agency's library doubles as an office for one of the staff attorneys. And the room where formal hearings take place barely holds fifty people, including the three commissioners, their staff and attorneys, and witnesses for opposing sides.

Despite the cramped space, the NHPUC has been the site of some of the hottest regulatory battles in New England. The controversy over the twin units of the Seabrook nuclear project, forty-five miles away, has focused a great deal of pressure on this agency. Its rulings have angered consumers, contributed to toppling conservative Governor Meldrim Thomson from office in 1978, and spawned a wide-ranging debate on how future power needs will be met in New England. It has also taught a number of people something about the politics of the power industry.

It was in the small New Hampshire hearing room that Mary Metcalf, a retired fourth-grade schoolteacher from Durham, first discovered the inner workings of regulation and how the lead builder of the Seabrook project, Public Service Company (PSC), operates. Her experience and impressions provide a vivid sense of the questions that have emerged over the will and capability of state regulatory commissions.

Although she was opposed to the construction of the nuclear reactors, Metcalf had never participated in demonstrations at the Seabrook site or lobbied in the state legislature against its construction. Instead she went to the state utility commission offices. The first hearing she attended in May 1979 concerned an $18 million rate increase and controversial bills for Seabrook's construction, known as construction-work-in-progress charges. She says that the first time she sat in the hearing room she was struck by the absence of other consumers. "These are public meetings, but there is no public," she says. "Many people aren't even aware that there is such a process." Other common criticisms include the fact that hearings are often held during working hours, receive little publicity, and are seen as the inner sanctum of lawyers, company executives, hired experts, and political appointees—"professionals."

Although Metcalf admits she was shy at first, she took notes and was determined to learn more. The hearings on the issue of construction-work-in-progress charges went on for nearly a year, severely straining New Hampshire's regulatory system. Five public meetings were held across the state and thirty-two hearings at the commission. The official transcript containing testimony and argument ran 5,000 pages.

Since that time, Mary Metcalf has become known as the New Hampshire Watchdog, steeping herself in the regulatory process by regularly attending hearings, examining thousands of pages of prefiled testimony, and discussing issues with consumer advocates and attorneys for whom she sometimes takes notes. Over that time, her impressions have not changed much. The regulatory process, she explains, "is not very well balanced nor sophisticated." Quite often, the state consumer advocate or the community action officer is the only person representing consumers' interests against a battery of company lawyers and hired experts. The consumer advocates, Metcalf says, "just rely on their common sense because they usually have no money to hire expert witnesses to speak on behalf of the consumer." She believes that the commission tries to balance the interests of the power companies against those of the general public, but the agency does not have the resources to weigh the information the company presents. No one on the staff is well enough versed in nuclear and financial issues to make judgments concerning how reasonable construction deadlines or methods of financing may be. "They ask good questions," Metcalf says, "but they aren't trained to evaluate the answers." The commission staff in New Hampshire, as in many other states, often relies on data generated by the power company's computer to analyze information presented during a hearing. The function of the commission at worst becomes a process of rubber-stamping the power company's plans, and at best, it issues orders on areas where it has expertise and stalls and muddles through what it can't understand while facing intense political and economic pressures. The result in New Hampshire has been disastrous.

In early 1984, overburdened with debt from massive cost escalations of the Seabrook nuclear project, PSC began to slide toward bankruptcy. Over the years, optimistic costs for the reactors had risen from $900 million to $9 billion. The completion date for the first unit had slid from 1979 to 1986, with an increasing number of financial observers doubting that the project would ever generate electricity. At the end of 1983, as funds ran perilously low, creditors and the project's sixteen joint owners grew anxious. Amid grave uncertainty, Robert Hildreth, a vice-president of Merrill Lynch, came forward with a bailout plan calling for creation of a new corporation to oversee construction and a billion dollars in additional financing. Merrill Lynch, as well as other prominent Wall Street firms, banks, and insurance companies, had a great deal of money tied up in PSC and the Seabrook project. The bailout consisted of three phases: raising an initial $90 million in capital and extending loans amounting to over $200 million; issuing bonds for $430 million; and creation of a new company to raise additional funds and oversee the project. It was welcomed as perhaps the only plan that would save the company and the project. The catch was that it would have to be approved not only by the joint owners of the project, but by the NHPUC as well. The process that followed showed the lack of sophistication that Mary Metcalf had witnessed for five years and also gave a glimpse of the raw politics at the center of the regulatory process.

In April 1984, the bailout plan was agreed to by the joint owners, and the first phase received approval from the NHPUC. Consumer advocates had argued that power from the project would not be needed until the mid-1990s and that cheaper methods of power generation existed. They pointed to a 1980 study by the Energy Systems Research Group of Boston, showing that a specific renewable energy program using hydropower, photovoltaics, wind, wood, and waste energy systems, put in place by the early 1990s, would produce more power than the Seabrook reactors at less cost and create thousands of new jobs in the process. But the state commission moved ahead.

The second phase was due to come up in August. In late June, however, Paul McQuade, chairman of the commission, was making the rounds to the business community in his campaign for governor, praising the attributes of the bailout plan. It was apparent that the remaining phases of the plan would not receive an objective hearing. On August 28, the second phase was approved in a 2–1 split vote with McQuade voting in favor. Hearings on the plan had been rushed, and consumer advocates complained to the state court, citing a speech by McQuade at the Portsmouth Chamber of Commerce supporting the bailout plan. The court took an extraordinary step, forcing McQuade to step down from the hearings and nullifying the approval.

PSC was stunned by the court's action. Money to pay off lenders to the first phase of the plan was needed from the second phase, which was scheduled to issue $435 million in bonds on September 30, just three weeks away. Company executives who were touring the country and meeting with influential investors were called back to New Hampshire for emergency conferences. Governor John Sununu was pressed to replace McQuade as rapidly as possible with a temporary commissioner. Over the weekend the selection process took place. By Monday September 11, John Nassikas, a New Hampshire native and former Federal Power Commissioner under Richard Nixon, had been nominated to fill the spot. Over the objections of consumer advocates and Dudley Dudley, a member of the governor's executive council, Nassikas was approved twenty-six hours later. From beginning to end, the whole process had taken five days.

Nassikas, a Washington, D.C., attorney, packed his bags for New Hampshire. In the meantime, the more critics learned, the angrier they became. Nassikas, as it turned out, was the treasurer of Americans for Energy Independence, an organization promoting national energy security, but especially development of coal and nuclear plants. In 1983, the group had received funds from the Committee for Energy Awareness (CEA), the industry organization running national pronuclear advertising campaigns. CEA had in turn received $266,635 from PSC. Additionally, Nassikas' law firm, Squire, Sanders and Dempsey, represented several power companies and General Electric, a turbine contractor for Seabrook. Fearing the worst, consumer advocates hoped that Nassikas would at least conduct a thorough hearing at which all the facts might be laid on the table. They were in for another disappointment, however.

A week after arriving in New Hampshire, Nassikas decided to forgo additional hearings. Basing his decision on his review of hearing documents, he cast his swing vote in favor of the plan. Kirk Stone, director of the New Hampshire Campaign for Ratepayer's Rights, said, "The record had been developed by McQuade in a hearing based primarily on testimony by a vice-president from Public Service Company and Hildreth of Merrill Lynch. The company came in with a gun to its head, saying that it would be bankrupt if decisions weren't made in time. Hearings that would ordinarily take a few months were being rushed through in two or three days. There was no time for intervenors to develop their case." Stone said the decision to allow the $435 million financing locked PSC into their plan and threatened to "double rates for many New England consumers in order to save Public Service Company."

The problems in New Hampshire were not isolated events, and the shortcomings that Metcalf ana Stone witnessed were not machinations apparent only to consumer advocates. Such conflict of interest and industry-orientation among regulatory commissions is an onerous tradition. Historically, most regulatory officials have shared the primary belief of the power empire that the nation's prosperity rises or falls on the availability of abundant supplies of cheap electricity. For these regulators, the job means helping industry expand. Michael Johnson, an outspoken member of the Pennsylvania Public Utilities Commission, notes the difficulties of bias in his state. He says, "The problems are within the structure itself; the problems are with the commissioners themselves. When you get people so beholden to an industry that their chief function is to carry out the wishes of that industry, you've got a problem of gigantic proportions." Appointed during the turbulent early 1970s, Johnson, a seventy-three-year-old former labor organizer, has fought his current fellow commissioners in what he calls "a real jungle—dominated by the utilities and the utility point of view." He also believes that the regulatory process is often no more than a rubber stamp for power company rate increases, which the companies "are trying to make more automatic."

Such candid views from a utility commissioner are rare, but it is generally accepted that political attitudes of state and federal officials are what shape regulation. Those who favor a "free market" and industry-oriented regulation have traditionally dominated state government. Their philosophy and policies to keep staffing and funding for regulatory agencies low have allowed the industry to run wild and build itself into a full-blown crisis. The $100 billion damage bill from cost overruns and cancellations of nuclear plants estimated by energy economist Charles Komanoff translates into rate increases for one third of the households in the nation. Industry executives now fear the backlash that is building over situations like Seabrook.

In Kansas, consumers faced a 95 percent rate increase from the Wolf Creek nuclear plant—the steepest proposed hike in the nation. Citizens and representatives of more than seven hundred businesses combined forces as the Alliance for Livable Electric Rates to stop the increase and its dire impact threatening the state's economy. The Kansas Corporation Commission received over 11,000 letters, petitions with 16,000 signatures, and witnessed meetings that brought out 3,400 people in Pittsburg and Wichita. In regulatory hearings the experts for the Alliance pointed out that the Wichita area alone would lose 7,000 to 13,000 jobs as electric bills rose. There were also strong threats to the local farm economy. "It's no secret that farms all over the state are really hurting," said Robert Eye, an attorney for the Alliance. "Any increase in overhead would be difficult to absorb, particularly for dairy operations that are dependent on electricity."

The power companies that took part in the Wolf Creek project "admitted that they had never considered any demographic data before considering to build the 1,150-megawatt plant," Eye said. Although the power companies claimed they would go bankrupt if they didn't get the full increase, the Kansas Corporation Commission slashed the proposed increase in rates by more than half for reasons of imprudence and excess capacity, and for the difference between the cost of Wolf Creek and the estimated cost of a new coal plant. The commission's tough stand came partly from new state legislation consumers had fought for and won in 1984 that allowed the cost comparison to an alternative power source.

In other states, consumers have fought for "cost caps" for plants. The cap limits the total plant costs and forces the company to absorb overruns. Caps have been used for the Limerick plant in Pennsylvania and the Millstone plant in Connecticut, but some consumer advocates charge that the caps are set at levels above what alternative power would cost and are aimed less at setting a bottom line for efficient management than squelching public protest over costs.

Peter Navarro, an economic researcher at Harvard University and a popular supporter of the power company viewpoint, warns that public pressure on regulatory commissions will undermine the financial viability of the power companies and will result in underbuilding of new power plants and a "dimming of America." Navarro points to Texas as a state where politicians have been elected on "no-rate-rise promises." He charges that the Texas regulatory commission has been "loaded with consumer advocates" and transformed from "one of the fairest and most technically competent in the country into one of the most rate suppressive."

Consumer advocates see the situation otherwise, praising the Texas commission for bringing a halt to overbuilding of power plants in the state and forcing the power companies to take advantage of the waste energy from the petrochemical industry—a cogeneration potential estimated to be equal to five new 1,000-megawatt power plants. Navarro also cites Massachusetts, Ohio, Florida, Mississippi, and New Mexico as being similarly consumer-oriented. Although Navarro claims to be an academic bystander in the conflict, Harvard's Energy and Environmental Policy Center with which he is associated received $10,000 in 1982 and $25,000 in 1983 from the Edison Electric Institute.

Another state Navarro could have added to the list is Montana, where the state regulatory commission shocked everyone by denying a 55 percent rate increase for the completed Colstrip 3 coal-fired plant in the summer of 1984. During the course of hearings on the rate increase, it became apparent that Montana Power Company was telling two different stories. A letter revealed that while the company had claimed the power was needed for Montana consumers, it had been negotiating to sell two thirds of its share under long-term contracts to California. In such a situation the company could increase rates in Montana to cover the costs of Colstrip, and make additional profits by selling most of its power to California utilities. Unfortunately, the decision was overturned in June 1985 by a district court, and a further rate hike fight over the new Colstrip 4 plant is expected.

□ □ □

Although regulation is the only way the public has to stop these schemes and control the giant monopolies that dominate the power industry, the regulatory field exists as a political no-man's-land. The conflicting mandate of the state regulatory agencies is to substitute for competition and provide protection for the interests of both consumers and the company. Power company executives express their understanding of this function as a "pact." The basic agreement, as they see it, is for the power companies to provide a steady supply of electricity and for the commissions to make sure that they get rates that make them highly profitable and able to borrow increasing amounts of funds from Wall Street for ever-larger construction programs. All of the major tax breaks and incentives they have lobbied for over the years are geared to support their expansion. And their rates are based on the value of their assets—as building expands, assets rise, and as assets rise, so do rates and profits.

Consumers have a different understanding. Their view is that power companies are given the privilege to operate as private monopolies and have a responsibility to keep their rates down by maintaining efficiency and promoting conservation—minimizing their growth rather than trying to maximize it. They consider tax breaks and government subsidies for the industry as "hidden costs" of electric power. They also demand that the "social costs," such as the danger of extra-high-voltage power lines, strip-mining, acid rain, radioactive waste dumps—or any of the array of broader impacts—be assessed when expansion is being considered.

The contest between these two views of regulation is what rages within the walls of government. It is an argument tied to philosophy of the roles of government and business in society, and its outcome can be measured in part by the amount of funding and staff that a state allocates to the job of regulation. Traditionally the consumer viewpoint and the scope of regulatory authority have been on the short end.

Peter Bradford, the chairman of the Maine Public Utility Commission in 1986 and a former member of the Nuclear Regulatory Commission, believes that one of the fundamental problems facing regulators today is the issue of having a competent staff that can investigate a wide range of issues in all of the necessary detail. Although regulatory commissions differ from one another in the nature of their statutory jurisdiction and powers, the overwhelming majority not only regulate electric, gas, telephone, and privately owned water companies, but also an array of railroads, common carriers, docks, wharves, warehouses, toll roads and bridges, canals, sewers, stockyards, and tunnels. One of the common denominators running through nearly all of the regulatory agencies is that, as in New Hampshire, the agencies lack the staff and financial resources to do the job that needs to be done.

The state regulatory commission staffs range in size from a high of 866 in California to a low of fifteen full-time employees in Hawaii. Nearly half of the commissions have fewer than a hundred full-time employees. Besides meager appropriations from state legislatures, the commissions raise part of their revenue from the regulated utilities, but the amount that can be assessed to help defray expenses is often quite low and limited in most states by statute. In Maine, the Public Utility Commission spends about one dollar for every $350 in utility revenues that are regulated. National figures are even lower.

Equally important to the size and funding for a staff is its attitude. A traditional issue is whether a regulatory staff has been "captured" by a utility. This means having an industry orientation, as in Pennsylvania, or an "incestuous" relationship in which regulatory staff, on retirement from the agency, go to work for a power company at double their previous pay, or have worked for the utility before coming to the agency. In the early 1970s, for example, the Colorado Public Utility Commission was found to have five former employees of Colorado Public Service Company on its staff. Their combined service for the power company was fifty-four years. Among them, the commission counsel had fifteen years and the executive secretary had twelve years. Despite purely aboveboard dealings, this can create a "buddy-buddy" atmosphere, in which officials may minimize conflict and ultimately have a difficult time distinguishing between public duty and private interests. Even when "incest" is not the issue, an institutionalized attitude can be developed by the staff.

Harvey Salgo, a former attorney for the Massachusetts Department of Public Utilities, says, "Those who stay tend to be the wrong kind of people. They are the career civil servants who settle themselves in and learn a tremendous amount but lose certain aggressiveness. They no longer look very hard at numbers or policies." Salgo's view is supported by those of experienced intervenors.

Charles Komanoff, a New York energy economist who predicted the financial decline of nuclear projects, often testifies before public utility commissions on behalf of consumers. From his standpoint there is an appalling amount of conservativism and ignorance among the staffs of public service commissions. Komanoff says, "I view the staff of any bureaucracy as terribly important. I think that they help set the tone about what's feasible, and from my experience it's absolutely disgraceful to see the level of ignorance and self-serving actions, and most importantly the outright emotional hostility they show toward intervenors, whether they represent consumers or environmentalists or commercial businesses."

Greg Palast, a utility rate analyst and former director of the New York State legislature's Commission on Science and Technology, also testifies for consumer groups around the country. His impression is similar to Komanoff's. "Most commissions," he says, "are willing to lie down and play dead. The utility companies don't cook the books in secret. They cook the books and hand the recipe to the commission and say: 'Here's what we did in technical mumbo jumbo.' The commission goes right along with this—there is little sympathy for consumer challenges even when they are done with expertise."

Expressing a disdain similar to Komanoff's, Palast says, "Most utility commissions are too dumb, too completely ignorant of the subject to understand what is going on, and when we try to explain to them, they still don't understand." He says the typical utility commissioner is a political crony who is found acceptable to the utility companies.

From his experience, Salgo says, conflicting mandates for the commissions and riptides in state politics often interfere with the positive attitudes of a regulatory agency's staff. The terms of commissioners, for example, usually run four to six years. In Massachusetts and a number of other states, a term of four years runs coterminous with the office of the governor. When there's a gubernatorial change, there is usually an extensive turnover in staff. Staff members who may have been encouraged by an aggressive commission can easily find themselves stifled under a new governor and a new set of commissioners, leading to demoralization and staff resignations.

This composition of the staff, its aggressiveness, and the time it can allocate to cases is critical in piercing the divisive arguments or false information that a company may provide. In a large number of cases, it's company whistle-blowers or consumer advocates who are able to unearth a misrepresentation of facts. In 1982, it was alleged that Northern Indiana Public Service Company (NIPSCO) had tried to collect double charges. Greg Palast was hired by the city of Gary, Indiana, to look into the expenses for the Bailly nuclear project. He testified that the company had charged customers $24.8 million in union wages, wages that had gone unpaid during an eight month strike in 1980. "The public was stuck paying for phantom workers," Palast says. In addition, according to Palast, the company charged extra overtime costs for payments to supervisors and nonunion workers during the strike and pocketed the savings. Palast says that in strike situations, such double-billing tends to be the rule rather than the exception. The Long Island Lighting Company, for example, laid off 1,000 workers to cut costs. The money saved did not go to reducing consumer bills, but to help pay for a $40 million monthly interest charge on the Shoreham nuclear plant, which was completed but remained inoperable because of a dispute over adequacy of emergency evacuation plans.

At the same time NIPSCO was trying to double-charge consumers for the Bailly nuclear project, a scandal broke out in Maine over false information given under oath by a vice-president of Central Maine Power, the state's largest utility. Maine Commissioner Peter Bradford says the commission draws a sharp line between a company trying "to blow expenses through that really don't fit the description that they hang on them" and "outright lying." "Within limits," he explains, "we expect a certain amount of that [blowing up expenses] and we try to knock it off when we can. But we usually don't penalize beyond disallowing it."

At a hearing in September 1982, Central Maine Senior Vice-President Robert Scott was asked about a company survey of public opinion on an antinuclear referendum and conservation programs. He told the commission that it had been destroyed when, in fact, it had not. Scott wanted to cover up a number of uses the survey had been put to, including identification of individuals who might help political candidates the company supported. According to an investigation by the state attorney general, Scott went back to his office after the hearing and ordered an employee to destroy the report. The company president, Elwin Thurlow, also ordered it erased from the company's computer. The employee didn't follow his orders, however. He drove around with the report in his car for two or three weeks and then turned it over to authorities and explained the situation. The Maine Commission subsequently cited Scott for contempt and fined the company $20,000. Ultimately both Scott and Thurlow resigned.

The most notorious regulatory scam in recent years occurred in New Mexico, where Public Service Company of New Mexico attempted to classify the cost of a new Lear jet as a construction cost. Ratepayers might still be unknowingly financing this corporate status symbol if it hadn't been for the efforts of an Albuquerque-based citizens group, Energy Consumers of New Mexico. Upon learning of the purchase, the group filed a complaint with the state public service commission asking that the cost be disallowed from consumer bills. As ludicrous as it seemed, the company did not give up without a fight, absorbing commission time and taxpayers' money.

The company attempted to show that use of the $3.3 million jet would save $4.1 million over a seven-year period. According to the company's only witness, assistant treasurer Albert Robinson, the jet would save the company for air freight on parts and passenger tickets. The commission discovered, however, that Robinson's $4.1 million figure was highly inflated. The New Mexico Commission called the purchase of the jet "an unwise and imprudent expenditure in light of its costs and benefits" and ruled that it couldn't be charged as a "construction" expense.

The above examples illustrate the diversity of the chicanery that goes on. A larger issue, however, affecting every state, and one that greatly contributed to the $100 billion damage bill of cost overruns and plant abandonments, is construction-work-in-progress (CWIP) charges. This practice, which allows power companies to automatically charge consumers for plants while they are being built, provided a blank check for industry construction programs during the 1970s. In many states consumers fought pitched legislative battles to have this policy overturned. In New Hampshire the political uproar over rising rates toppled conservative Governor Meldrim Thomson from office when he took a stand defending CWIP charges and a series of rate increases for the Seabrook project.

In addition to CWIP, a broad array of questionable charges are passed through to consumers. One is "phantom taxes." Since World War II, a clause in the federal tax code has allowed private power companies a special accounting treatment in which taxes can be deferred and placed in special accounts. By the time these taxes come due, the power companies receive new tax deferrals to offset them. Meanwhile, the companies invest the accumulated funds. The $1.8 billion total of "phantom taxes" noted in 1967 by Vic Reinemer and Senator Lee Metcalf in the book *Overcharge* had grown to more than $42 billion by 1985.

Another charge is for dues paid to Edison Electric Institute, which can then use the funds to lobby against consumer interests. In 1985, the National Association of Regulatory Utility Commissioners was investigating the use of some $25 million of EEI's funds spent in 1984. Some states have struck a portion of EEI dues from consumer bills, but the primary portion of the organization's funding is in consumer bills.

Another charge passed to consumers is for advertising, which, ironically enough, often includes costs to persuade the public that nuclear power is necessary. Connecticut Light and Power, for example, spent $962,457 in 1983 on publicity and advertising, largely to promote coal and nuclear plants. Its affiliate, Western Massachusetts Electric Company, paid $173,142. Louisiana Power and Light paid the Committee for Energy Awareness $364,155 for nuclear advertising and charged it as an "operating expense" listing "national welfare" as the purpose of the expense. New Orleans Public Service Company paid $19,087 under an identical explanation. The Middle South holding company, through all of its subsidiaries, contributed over $1 million. These contributions are standard practice for nearly every company.

The deepest irony of all is in charges added for rate increase cases. While regulatory commissions and state consumer counsels are understaffed and underfunded, power companies can add their costs for attorneys and expert witnesses to rate increases—forcing consumers to pay for the case against their own interests. This lopsided allocation of resources assures power companies of being able to put on a stronger case.

Outside the regulatory institution are hundreds of grass-roots consumer-advocacy groups that operate in virtually every state, which attempt to speak for consumer interests. The groups represent a variety of constituents and are often more aggressive than state consumer counsels working in commission or attorney general offices. However, as nonprofit organizations, they rely largely on donated services and the rare experts who are willing to speak out against power companies. Organizations such as California's Toward Utility Rate Normalization, New Hampshire Campaign for Ratepayer Rights, Illinois Public Action Council, Wisconsin's Environmental Decade, and the Southwest Resource Center have provided landmark testimony and helped to lead consumers in a growing movement for regulatory reform.

Along with shortages of funding, these groups often face the political hostility Komanoff describes and the inertia and bias of regulatory institutions. On construction project issues, for example, the burden of proof is often shifted to the consumer to show why a power company's plans are wrong, instead of the company proving that it is right. This is especially true in the case of plants under construction, which were considered reasonable in the past but are no longer prudent. The job requires a tremendous amount of time, expertise, and consultation with experts in legal and economic fields. In some cases, awards of state-controlled "intervenor funding" are available to help consumers develop a case, but such awards are extremely limited, and winning one is the occasion for celebration.

Despite the deeply political nature of regulation and the lopsided balance of resources, consumer groups have fought tenaciously to save consumers tens of billions of dollars. Additionally, there are also some enlightened state regulatory commissions, notably in Wisconsin, Maine, and California, which value consumer input and are willing to consider innovations. For the most part, however, the commissions are caught in the web of state politics and a belief that their job is to help promote maximum expansion of energy consumption. The most cynical view is that they were never intended to regulate power companies fully, and that, long before the current crises, deep-seated problems were visible.

□ □ □

Many critics agree with Michael Johnson's assessment that the real problem is in the institution of regulation itself. Since most industrialized countries have power systems that are predominantly publicly owned, the extensive state regulatory system is peculiar to the United States. Wrapped in back-room politics and established with the conflicting mandate of protecting both consumers and the companies, regulation is an inherently impossible task—or so some observers believe.

The current failure of regulation is evidenced on the one hand by the

$100 billion damage bill of nuclear power and on the other by the power industry's threat that we may see a "dimming of America." Over time, the process of power company regulation has gone through four distinct phases, in which efforts at public control of the giant private power monopolies have failed and engendered new waves of bureaucratic fixes and reforms: 1879–1906; 1907–1934; 1935–1978; 1979– .

In the first phase of regulation under a local franchise, the early private power companies became embroiled in ruthless competition and municipal corruption. This gave support to proposals for the companies to have a natural monopoly under state regulation. As detailed earlier, the drive for state regulation came from the private power companies' desire to head off local public ownership of the industry by consumers. Previously, laws for the regulation of railroad rates had been passed as an alternative to public ownership of the railroads, which the Grange Association and others were advocating. The law for power company regulation was based on a railroad case known as *Smith* v. *Ames,* which established the right to charge rates based on the fair value of assets. But there was an open interpretation as to what constituted "fair" value. Faced with the surge toward municipal ownership of electric companies in the 1890s and early 1900s, Samuel Insull and his colleagues came forward with plans for state regulation.

Although the idea of state regulation was strongly criticized by Milwaukee Mayor Daniel Hoan and others, it was supported by the influential National Civic Federation, the National Electric Light Association, and a number of leading industrialists. The second phase of regulation began with the establishment of state regulatory agencies in Wisconsin and New York in 1907. Problems continued, however. The fundamental difficulty was that the mandate to protect both consumers and the interests of the companies was so broad and contradictory as to be nearly meaningless. The initial legislation provided no measures or guidelines for making decisions; thus any rulings were subjected to supercontrol by the courts. The courts' orientation was to see rights and policies more from the standpoint of property and invested capital than that of common interests. Thus from the start even those hopeful for the good purposes and intentions of state regulation watched as the system was undermined. Coupled with a political atmosphere that encouraged the supremacy of corporate interests, regulation evolved into meetings in which the commission and company negotiated rates up or down. There was no thorough investigation into how bloated the assets of a company might be or the wisdom of disparate expansion. With the basic assumptions of law focused on property and the perception of power companies as natural monopolies, the process of regulation was narrowed to a minimal maintenance function and doomed from the start. In *Electric Utilities: The Crisis in Public Control,* William Mosher wrote, "Ratemaking in 1928, despite a score of years of effort on the part of regulatory bodies, is as hopelessly muddled, indefinite and unscientific as it has ever been." It was clear in the unrestrained machinations of the holding companies that state regulatory agencies were not the answer to public control over the industry.

At the national level too, regulation was literally a hollow institution. In 1920, the federal government entered the field of electric power regulation with passage of the Federal Water Power Act. The act was the final compromise in the battle over control of the nation's rivers that Teddy Roosevelt and Gifford Pinchot had attempted to bring to a head years before. Although there was great concern over wholesale power being sold across state lines and escaping state regulation, the new commission's authority was limited to the siting and financing of dams. For its first ten years, the agency was actually nothing more than an interdepartmental coordinating committee made up of the Secretary of War, the Secretary of the Interior, and the Secretary of Commerce. It had no staff of its own, but borrowed from the three departments. The three secretaries spent an average of five hours a year on Federal Power Commission (FPC) business.

Given the sympathies of presidents Harding, Coolidge, and Hoover for private power development, it is no surprise the FPC was little more than a licensing bureau in the 1920s. During the first three years of its existence, twice as much generating capacity was being built as had been installed under permits from various departments during the preceding twenty years. The work of the FPC's borrowed staff was often hampered by a refusal of private companies to comply with information requests. At a Senate hearing held in 1930, the commission's chief accountant said that many companies had failed to comply with commission requirements for information, abide by regulations, or file statements. Several companies also refused to give the commission staff access to their records. During its first ten years, the commission had never once relied on the basic investment provisions to measure hydroelectric rates charged by the companies. With limited staff resources, it was decided that "rather than risk litigation it would be better to let sleeping dogs lie and seek agreements through regulation." At the Senate hearings, Representative Emanuel Celler of New York observed, "Congress might just as well have put the King of England, Mussolini, and Albert Einstein on the commission so far as any spontaneous decisive action is concerned."

Congress reorganized the FPC in 1930 under a new Federal Water Power Act. But significantly, the new commission, consisting of five full-time members assisted by a technical staff, still had no jurisdiction over the sale of electricity over state lines by the holding companies, which generated 85 percent of the nation's power. It was the collapse of the holding company empires and the revelations of the four-year-long Federal Trade Commission investigation on abuses in power company lobbying, business practices, and propaganda that finally sparked a third phase of regulation under Roosevelt's New Deal.

Roosevelt tried to provide expanded authority to regulatory agencies and reinforced this with the yardstick of public power against which the operations of private companies could be measured. These efforts were only partially successful. In 1935, Congress again amended the Federal Power Act with new provisions to allow for oversight of wholesale power contracts and to separate the relationship of Wall Street and the private power companies by prohibiting utility executives or directors from holding positions with another utility or banking firm without specific commission approval. Shortly after this provision became effective, the commission received more than eight hundred applications. They got prompt approval. This process continued into the 1980s with as many as 1,000 directors of power companies also sitting on the boards of banks.

The Public Utility Holding Company (PUHC) Act of 1935, which sought to break up holding company empires, was also loaded with loopholes. Nine major electric holding companies and seventy or so smaller electric holding companies were allowed to continue to operate. Coupled with the exemptions under the Federal Power Act, there was not a full restructure of the industry or a full extension of public control. The core of the old abuses and concentration of corporate dealing, though less direct, remained intact as shown by the Dixon-Yates investigation in 1955.

At the state level, where regulation had clearly failed, Roosevelt hoped that strengthening public power systems and the ability of citizens to buy out private power companies would force the private companies to shape up. By the 1950s, however, the private power interests, backed by the Eisenhower Administration, had decimated the yardstick concept by equating public power with communism and undermining federal power programs from which many of the public systems received their electric supply. There were a few regulatory analysts at that time who realized what failure of the yardstick and restructure of the industry might mean.

In his book, *Transforming Public Utility Regulation,* published in 1950, John Bauer, a former lecturer in public utility regulation at Cornell, Princeton, and Columbia universities, warned that if regulation could not be reformed, there would again come a clamor for outright public ownership. While the industry had grown more complex in the wake of the New Deal, corresponding regulatory evolution had been blocked. Bauer wrote, "The basic trouble has been a lack of positive legislative measures designed for equal protection of investors and consumers, for advancement of efficiency and economy, and for aggressive promotion of the public interest." He proposed model legislation providing greater coordination between federal and state agencies and a systematic administration of clearly defined standards that would move toward what would later be known in the 1980s as least-cost policies.

In regard to the companies themselves, Bauer urged that the public service aspect of their operations be recognized by placing consumers, as well as representatives of labor and each class of stockholders, on the board of directors. Rather than the mix of common and preferred stock and bonds used for financing, Bauer recommended fixed-income securities that would take the control of power companies out of the hands of speculators.

Unfortunately his warnings and proposals were not heeded. Under the Eisenhower Administration, regulatory agencies at the federal and state levels remained highly politicized bureaucracies, maintaining a narrow focus on rates and encouraging the build-and-grow mentality of the power companies. No effort was made to look ahead and help shape future plans of power companies and their growth impact on communities and the environment. As in the infamous examples of the AEC's involvement with the Fermi breeder reactor in Michigan and the massive subsidization of Shippingport and other reactors, the political orientation of regulatory agencies was not to guard the interests of consumers, but to help build and expand the industry. This growing failure of regulation set the atmosphere for ongoing turbulence.

In the early 1960s regulation was shaped by the rumors of an approaching fossil fuel shortage and the need to keep rates down as power company building programs began to expand. After 1965, however, rates began to climb steadily for the first time in the industry's history. In 1967 Senator Lee Metcalf told a congressional committee that electric utility regulation had deteriorated to the point that it was meaningless in many states. "Weak and toothless laws," Metcalf declared, "permit the public to be overcharged hundreds of millions of dollars."

As rates continued to rise, local and statewide consumer groups began to form and intervene in regulatory hearings. Environmental and rate issues became quickly linked. It was obvious to environmentalists that they were not fighting just the siting of single plants, but larger issues of power expansion and consumption. They began to question why so much power was needed and if it couldn't be produced and used more efficiently. One obvious problem was the way charges were set up in a "declining block" system, which encouraged consumption by offering increasingly cheaper rates as greater volumes of power were used. It might have helped to sell electricity, but it discouraged conservation and efficient use of power. In 1971, in a landmark case before the Wisconsin Public Service Commission, the Environmental Defense Fund, a national public interest group, successfully proposed a different system for "time of day" rates. Under the new system customers were charged more for using power during peak hours and less at times when demand was low. This flattened the demand power companies had to meet and helped eliminate the need for at least a few new plants in each area. It was the beginning of a broader discussion over rates stirring in the outer chambers of regulatory institutions.

When the oil embargo hit in the fall of 1973, all hell broke loose. The electric power industry was dependent on oil for 17 percent of its generation nationwide. But in some regions such as the Northeast, generating plants were 60 percent oil-fired. As OPEC producers cut shipments to the United States in a dispute over American support of Israel in the Arab-Israeli war, oil prices rose eightfold, and electricity prices went up by nearly 50 percent.

Power companies passed these soaring charges through to consumers in the form of a "fuel adjustment charge," a billing procedure that had been established during the First World War. Consumers, already angered by rising electric bills, turned to the fledgling consumer organizations for help and tried to keep their bills down by shutting off lights, turning down thermostats, and using appliances less. This resulted in a second shock for the power companies. In 1974, for the first time since 1946, the steady growth in the demand for electricity declined. This event pointed to a fact that both the power companies and regulators refused to recognize: The nation's power use was vastly inflated, and many of the new power plants proposed or being built were not needed. The immediate impact of the shock went deep, affecting not only consumers but investors as well.

In New York, Consolidated Edison, dependent on oil for 75 percent of its power, was hit especially hard by rising costs and a drop in the growth of power demand. Before 1973 the company had paid $4 to $5 for a barrel of oil; the price was now $23 or $24. Although New York Public Service Commission gave the company a temporary rate increase of $75 million in February 1974, by April cash reserves were so low that the company was forced to omit paying dividends on its common stock. This action, which was the first dividend omission in the eighty-nine-year history of the giant company, hit the industry with the impact of a wrecking ball. It smashed investors' confidence and sent a shock reverberating through the industry. That same month, the price for the average utility stock fell 18 percent; by September, utility stock prices were off by 36 percent. Nationwide, regulators were pressured by Wall Street and power companies to allow still higher charges to consumers so that the damage to investors' confidence could be repaired.

Regulatory commissions, which had led relatively quiet existences in the backwaters of state government, found themselves ill-equipped to handle the blast of attention suddenly focused on them. In the postembargo period, the number of rate cases increased more than 500 percent as utility executives requested a series of larger and larger rate increases. The state commissions were snowed under with work. Annual appropriations and staff for the regulatory agencies were enlarged, but with few exceptions, notably New York and California, the additional economic and technical resources made available were insufficient to keep pace with the ever-expanding case load.

Nearly all public utility commissions added automatic fuel adjustment clauses to their regulations, designed to pass increased costs directly to consumers without the benefit of public hearings. In 1975, nearly $5.9 billion in fuel adjustment increases were passed to consumers. As prices soared, so did consumer anger.

In 1974 and 1975 local and statewide consumer groups came forward, pressing for institutional changes in regulation. They organized large, angry groups of people to attend hearings. It was opposition the power companies had never seen before. At the state level, consumer groups took to shopping malls and street corners with petition drives to outlaw automatic fuel adjustment charges. Fuel adjustment charges without hearings, they argued, deprived consumers of due process and effectively hid the true fuel costs from the public. In several states the legislatures passed laws requiring commissions to conduct monthly audits of utilities, semiannual hearings on fuel charges, and annual review of power company procurement policies.

Consumer advocates also proposed an array of changes in the structure of electric rates. Along with tipping the declining block upside down to charge less for low power users and more for larger volumes, they fought for a special "lifeline" rate for the elderly and those on fixed incomes and time-of-day rates. They also attacked phantom taxes and construction-work-in-progress charges. One utility executive said that the irrepressible movement for rate reform was like a brushfire: "As soon as you stamp out one outbreak, it pops up somewhere else."

The federal government responded to the uproar by reorganizing agencies and shuffling personnel in the Energy Reorganization Act of 1974. But there were no substantive changes in policies. One new agency organized in 1974, the Federal Energy Administration, was to oversee development and implementation of petroleum regulation and allocation programs. It exemplified the lack of substantial change and continuing orientation toward industry. Its staff found gross inflation of prices in "daisy-chaining" schemes, in which a ring of suppliers would trade a shipment of oil back and forth on paper, inflating its costs with each transaction. When news of the practice began to surface, the investigation was stifled and staff transferred. Eventually the agency would become part of a new Department of Energy.

In that same shuffle, the Atomic Energy Commission, discredited by its efforts to both promote and regulate nuclear power, was split into two agencies: the Energy Research and Development Administration (ERDA) and the Nuclear Regulatory Commission (NRC). Although ERDA was mandated to explore all forms of energy, four fifths of its staff was from the AEC and their focus was on centralized, capital-intensive energy technology. John M. Teem, who served as ERDA's assistant administrator for solar and geothermal and advance energy systems, resigned in January 1976, charging the Ford Administration with not giving solar energy the priority it deserved.

It was obvious that the crisis in the third phase of regulation had to give way to something new. Believing power company claims that "shortages and unreliable supplies of electricity were imminent" the Ford Administration attempted to bail out the power industry and accelerate nuclear construction with the Utility Reorganization Act of 1975. Among other things the act called for automatic fuel adjustment charges and construction-work-in-progress charges to be reinstated in states that had knocked them out. Peter Bradford and other critics of the act testified before a Senate subcommittee that it was bad law, bad economic policy, and especially bad and unimaginative energy policy. Consumer advocates and power industry lobbyists managed to deadlock Congress over this issue, and the act failed to pass.

In the presidential election of 1976, energy was a central issue. Shortly after his election, President Jimmy Carter proposed a much debated National Energy Plan centered on conservation programs. The power industry fought fiercely against the bill and defeated it. But out of its ashes rose the National Energy Act of 1978. It contained a deal that split the opposition of energy producers by offering natural gas companies deregulation of new gas wells in turn for a measure known as the Public Utility Regulatory Policy Act (PURPA) of 1978, which promised to have a significant impact on conservation and renewable energy.

PURPA provided a potpourri of "nuts and bolts" reform and conservation guidelines to be undertaken at the state level. In its pure form, the law posed a potential revolution in the power industry. It required regulatory agencies in each state to compile a series of standards to promote conservation, energy efficiency, and equity in utility policies by November 1981. The law also required power companies to purchase electricity produced by owners of small dams, windmills, and industrial facilities and limited the company's option to take over such independent ventures to a 50 percent ownership. This would reverse the age-old trend of centralization and bring a wave of new economic competition to the natural monopoly the industry had enjoyed for so long. Although PURPA had none of the fanfare of the New Deal legislation, it was as far-reaching as the laws that had attempted to reorganize the industry in the 1930s. The problem was implementing it.

□ □ □

For five years, the power companies tied up the Public Utility Regulatory Policies Act (PURPA) in the courts. A number of states, although hampered by power company resistance, worked to enact its provisions. Others waited to see what the outcome of the legal fights would be. The most serious challenge came from the Mississippi Power and Light Company, part of the Middle South holding company system. With little surprise to local consumer advocates, the company was joined in its suit by the Mississippi Public Utility Commission, which sought to prevent Congress from providing guidelines for state rate-making considerations, or conservation and alternative energy development. In February 1981, U.S. District Court Judge Harold Cox agreed with the power company and initially struck down the law, saying, "The sovereign state of Mississippi is not a robot or lackey which may be shuttled back and forth to suit the whims and caprice of the federal government. . . ."

The following year, another U.S. district court judge in Washington, D.C., invalidated rules under which power companies would be required to purchase electricity generated by independent owners of dams or other energy facilities. That suit had been brought by the American Electric Power holding company system, Con Ed of New York, and the Colorado-Ute Electric Association. While the Mississippi ruling declared PURPA unconstitutional, the Washington, D.C., ruling knocked the bottom out of its provisions to encourage independent power production.

Finally the case went before the Supreme Court. In a close 5–4 decision, the Supreme Court upheld the constitutionality of the law and specifically the provisions requiring utilities to purchase power from independent sources. The decision marked a huge victory for consumers. Voting in the majority opinion, Justice Harry Blackmun wrote, "It is difficult to conceive of a more basic element of interstate commerce than electric energy, a product used in virtually every home and every commercial or manufacturing facility." Justice Sandra Day O'Connor disagreed with the decision, stating that PURPA "permits Congress to kidnap state utility commissions into the national regulatory family." Although implementation of the provisions would begin on paper, state commissions continued to drag their feet, and power companies were by no means at the point of giving up their resistance. Within the new Reagan Administration, efforts were also afoot to undermine the law.

While the legal battle over PURPA had been building, the Reagan Administration arrived in Washington with a vision of dismantling federal regulatory agencies and removing federal oversight of the power industry. The Reagan Administration believed that power companies and their millions of customers were victims of federal energy policies that diminished domestic energy production and forced dependency on expensive foreign oil. Reagan analysts blamed "overregulation" for driving up electric rates, adding to our balance of payment deficits and weakening the dollar.

The Reagan policies were formulated and promoted by a host of individuals with industry backgrounds. In addition to President Reagan, who had served as a spokesperson for GE for nearly a decade and promoted Senator Barry Goldwater's stance against public power in 1964, the executive level of the administration included three officials who had been employed by the Bechtel Corporation, one of the largest power project constructors in the world. George Shultz, Secretary of State, had been president of Bechtel. W. Kenneth Davis, deputy director at the Department of Energy, had been a Bechtel vice-president. Caspar Weinberger, Secretary of Defense, was a former Bechtel lawyer.

The head of the Federal Energy Regulatory Commission (FERC) was C. Michael Butler III, an advocate for the power industry for over a decade. As senior attorney for the American Natural Resource Council of Detroit between 1976 and 1979, Butler had worked to arrange loan guarantees and financing for the Great Plains Coal Project in North Dakota. Later, when he worked for Senator John Tower of Texas, he was a consistent supporter of legislation on nuclear waste, limiting the windfall profits tax on oil and resource development on federal land. Most significantly, he helped develop the 1980 Republican Party platform calling for deregulation of electricity and gas production and de-emphasis on funding conservation and renewable energy technologies. In November 1983 Butler was replaced by Raymond J. O'Connor, an executive vice-president of Prudential-Bache Securities Inc., who launched a number of initiatives to loosen government control of the industry and deregulate electricity and gas markets.

The Department of Energy under Ronald Reagan was at first headed by James Edwards, a former South Carolina governor and dentist, who proclaimed his mission was to work himself out of a job by dismantling the agency. Edwards was an unabashed supporter of nuclear power. He called South Carolina the "nuclear capital of the world." In testimony before a House subcommittee in 1978, he was asked whether South Carolina residents wanted a nuclear waste repository in their state. He responded that South Carolina already was a nuclear waste repository. "I make no excuses for it. I'm not apologizing for it. Thank God we have these nuclear wastes, because if we didn't we would probably all be working in salt mines in Siberia today, and we would not be worried about nuclear wastes," he said.

The Reagan energy plan called for "efficient energy production." In the short term this included reforming the federal regulatory process to remove impediments to the use of coal, speeding up the nuclear licensing process, and working with state regulatory bodies to streamline changes in siting of plants and nuclear waste disposal. Subsidies for nuclear power were boosted to more than a billion dollars annually while the solar budget was slashed 87 percent. An attempt was also made to eliminate virtually every federal energy conservation program, including those for insulating schools, fire stations, and other public buildings. Ironically, despite the massive subsidy for nuclear power, Reagan officials claimed that "free market forces" and consumers would make the final judgment over energy alternatives.

In one critical controversy, allegations emerged that Reagan Administration officials sought to undermine the independence of the Office of Investigations at the Nuclear Regulatory Commission. In April 1986, Julian Greenspun, who resigned after more than six years as the chief prosecutor of NRC criminal cases at the Justice Department, said NRC commissioners and top officials "don't want criminal cases brought, and they're willing to do almost anything to see that that doesn't occur. . . . Watergate hasn't taught these people a thing. They're doing everything they can to neutralize the Office of Investigations." Commissioner James Asselstine candidly agreed that the NRC "has had a lot of trouble dealing with wrongdoing. . . . There has been too much of a closeness with industry." The ability of the NRC to investigate wrongdoing is critical in view of safey-problem cover-ups and the potential for disaster at nuclear plants.

In 1984 there were more than five thousand reportable incidents at nuclear plants in the United States. "The list of close calls strongly suggests that a major U.S. nuclear accident could be lurking around the corner," noted Congressman Ed Markey, who chairs the House subcommittee overseeing the NRC. In one event concerning Detroit Edison's new Fermi reactor, high-level utility officials were charged in a preliminary report with possibly misleading the commission on a mishap involving a premature chain reaction at the plant. The case was blocked from being referred to the Justice Department. Walter McCarthy, Detroit Edison's chief executive officer, later apologized to the NRC. And in a potentially dangerous incident concerning American Electric Power Company's D. C. Cook nuclear plant, officials assured the NRC in writing that the facility was complying with federal fire and safety regulations. NRC inspectors later learned that the company had not made millions of dollars in required improvements at the plant. At the time, the industry was mounting an unsuccessful legal challenge to overturn the NRC's fire safety rules.

The most celebrated case concerns the Three Mile Island accident. The Nuclear Regulatory Commission took no action for three years on allegations that workers at the plant had rigged important tests of leaks in the cooling-water system before the mishap. In 1986, the commission was still holding hearings on Metropolitan Edison Company's conduct—four years after it pleaded no contest to criminal charges that inadequate and falsified testimony had been used. Among 1985's most serious safety problems was the loss of feedwater systems at the Davis-Besse plant near Toledo and a 26-minute loss of power at the Rancho Seco facility near Sacramento, both requiring scrupulous attention.

In early 1986 the NRC's 23-person investigative staff was said to be struggling with a backlog of nearly two hundred pending cases. At a congressional hearing Commissioner Lando Zech, Jr., dismissed pleas to increase the size of the staff. Former Justice Department prosecutor

Greenspun told the *Washington Post* that the commission's shortcomings stemmed from "a well-meaning desire to see nuclear power succeed."

Some observers likened the Reagan Administration to that of Herbert Hoover in its firm commitment to the interests of large corporations. Michael Pertschuk, chairman of the FTC during the Carter years, saw the Reagan positions as more extreme than Hoover's. In an interview in October 1984, Pertschuk said that the Reagan policies did not represent "one pole in the normal pendulum swing of American politics. It is an aberration," he said. "For better or worse we have never before had regulatory agencies manned by those who loathe them and scorn their mission." For the power industry, such an attitude brought a mixed blessing. Deregulation of the electric industry was their ultimate aim, but it promised potential chaos if it progressed too rapidly, not only in reorganizing the industry into generating companies, transmission companies, and distribution companies, but in a backlash of anger from consumers.

In essence the lax atmosphere of regulation under Reagan officials helped deepen the growing crisis. By the early 1980s, weak industry-oriented regulation and a refusal to acknowledge the need for cutbacks in building programs created economic disasters long predicted by consumer and alternative energy advocates. In early 1982, rising costs of power and a sluggish economy combined to create a drop in the peak demand for electricity. Since 1974, the growth in electric use had slowed from 7 percent to 2 percent per year, but this was the first time there had been a drop in the peak demand since the Depression. The event marked a dramatic turning point for the industry, which had kept building new plants with the approval of regulators despite the decade-long slowdown. Against all reason, industry officials had convinced regulators that the power was needed. Then, as construction costs continued to climb, record-breaking rate increases had been allowed to keep the building going. In 1981 regulators allowed increases totaling $8.3 billion, quadruple the annual $2 billion to $3 billion increases of the 1970s. The total represented an average of 80 percent of what the industry requested. In 1982 rate increases totaled $7.6 billion. But then came the drop in peak demand. As new plants continued to come on line, the surplus of peak capacity reached 39 percent nationwide, more than double the recommended amount of 15 percent.

The citizen protests that hit the Northwest over high rates and surplus power in 1982 erupted in nearly every other state where new power plants were under construction and promised to bring rate increases of

30 to 300 percent. Analysts began talking of "death spirals" for power companies—a situation in which rates would be raised to meet fixed expenses, consumers would cut back on power use to bring down their bills, and rates would be raised yet higher, continuing in an ascending spiral until consumers revolted or the company went bankrupt because regulators called a halt to the increases. In many states regulators, faced with record surpluses and record rate increases to pay for more plants, began to talk of recommending cancellation. While some 250 coal and nuclear plants had been canceled since 1974, most had been in the planning stages. The plants considered for cancellation now stood as much as 85 percent complete and had billions of dollars already invested in their construction. With nothing but trouble in the wind, the power industry and Wall Street firms that stood behind the companies went to the Reagan administration for assurances that something could be done.

In February 1982, fifteen representatives of the power industry and Wall Street met with Vice-President George Bush at the White House to formulate a plan for increased financial support and an extensive loosening of regulation. The group was led by a self-labeled "gang of four," which included Stephen Bechtel, Jr., chairman of the Bechtel Corporation, Robert Kirby, the chairman of Westinghouse, Larry Wallace, the president-elect of the National Association of Regulatory Utility Commissioners, and Sherwood Smith, Jr., chairman of Carolina Power and Light and the head of EEI's Governmental Affairs Committee.

The administration agreed to establish a special task force to be chaired by J. Hunter Chiles III, the Department of Energy's director of policy planning and analysis. (Chiles was to leave the Department of Energy a year later to take a job with Bechtel Corporation.) With $17 million in funding, the Working Group on Electric Utilities set out to explore a variety of reform proposals, including deregulation of electrical generation, establishment of regional generating companies subject only to regulation by the FERC, possibilities for regional regulation, and other model policies that would be proposed for adoption by the states. These plans had the potential to undermine PURPA, speed centralization of the industry, and weaken public control.

As the power companies were pushing for resolution of the building debt crisis, the states were erupting in protest over soaring utility rates. Increasingly, consumer groups attempted to shift the control of regulation from the courts and back-room politics to the ballot box. In West Virginia, consumers, stung by natural gas prices that had increased 129 percent and electric prices that had soared 64 percent in five years, mounted a ten-month lobbying campaign. In the fall of 1982 several hundred people rallied at the Charlestown Civic Center, telling utility commissioners, "This is a democracy, people do rule, and we will use that." In November they succeeded in electing twenty-three of twenty-five consumer-backed candidates to the state legislature. The result was "a whole new ball game" in the legislature, according to David Grubb, who headed the state's Consumer Action Group. A bill for least-cost planning and comprehensive rate reform passed in March 1983. The legislation had taken its core from the federal PURPA legislation. Governor Jay Rockefeller called it, "The most significant piece of landmark legislation ever passed by the West Virginia legislature."

Similar success came in Iowa that same month. While Skip Laitner, director of the Community Action Research Group, crowed over "victory after victory," spokesmen for Iowa utilities called a legislative package passed there "devastating." The new law promised sweeping changes in utility charges, and as in West Virginia, it mandated least-cost planning for future growth. Consumers in Iowa were paying for a surplus of electricity that had reached 52 percent. One provision of the new law would force power companies to absorb costs for any excess capacity over 15 percent.

In as many as seventeen other states, consumer groups pushed for elected rather than appointed utility commissions. The Illinois Public Action Council published a study in 1981, pointing out that electric rates were lower in states with elected commissions and increased at a slower pace than in states with appointed commissions. Power industry supporters have attempted to dismiss this fact, claiming that in many states with elected commissions there is easy access to cheap fuel such as coal. While it explains part of the lower rates, elected commissions are oftentimes more aggressive in protecting consumers. Wall Street analysts consider the existence of an elected commission in any state a primary reason for downgrading a power company's stock.

Most utility reformers recognize that elected commissions are not a panacea. The Oklahoma and Alabama commissions are often pointed to as examples of where "good ole boys" have been elected and continue to do the bidding of the power companies. The commission in Alabama has the distinction of having had its chairman found guilty of soliciting a $10,000 bribe in 1971. Most of the proposals for elected commissions therefore contain provisions to protect and enhance accountability by restricting campaign contributions, limiting the number of terms a commissioner can serve, and prohibiting the "incestuous" relationships and revolving-door practice of taking industry jobs upon leaving the commission.

Another popular consumer strategy to strengthen consumer intervention is through the creation of statewide Citizen Utility Boards (CUBs). The idea for CUBs emerged from Ralph Nader's staff as a method of organizing a broad-based membership though flyers inserted with a consumer's utility bill. The first CUB was established by an act of the Wisconsin legislature in 1979. The law required the utilities to allow fund-raising appeals by the group in monthly billings. The Wisconsin CUB appeals were a shocking surprise for consumers, who opened the power company envelope to find another envelope inside saying, "Read this before you pay your bill." The note on the reverse side of the envelope severely criticized the power company and asked for a donation to help fight rate increases. The idea for CUBs has been under active consideration in more than a dozen states. In 1983 a second CUB came into being in Illinois and a third in Oregon in 1984. Other proposals were being debated in Florida, Massachusetts, Maine, New York, Rhode Island, and California. In 1985 a Supreme Court ruling on a California case declared that states may not force private power companies to include consumer-group inserts in their customer billing. Many observers believe that without being able to piggyback membership and fundraising, the CUB idea may wither.

While the CUBs and other organizations can pool and concentrate the limited resources of consumers, they do not address the fundamental issues of public or private control and are left to operate in a regulatory system largely designed and dominated by private utilities and their investors. For example, even if a consumer group is successful in keeping rates down, a direct response will be a downgrading of the power company's securities on Wall Street. This creates a higher cost for capital, which is automatically passed along in the form of higher rates. If a consumer group is extremely successful, a power company may take operational shortcuts and reduce personnel, thus reducing the level of service, in order to maintain profits.

The basic problem, of course, is that legislative packages mandating conservation, elected commissions, and other reform efforts are bureaucratic fixes for an institution that is highly influenced by politics and offers more protection to the interests of the industry than to consumers. Even if consumer groups succeed in creating laws for reform, such as PURPA or the legislation in Iowa or West Virginia, they still have an enormous task in consistently matching industry's funds and influence. The chief myth is that there is some point of equilibrium, where all sides are happy with costs and profits. The alternative myth is that the process of regulating large private companies that are the most capital intensive in the world can be depoliticized.

During the summer of 1984, the National Association of Regulatory Commissioners passed a resolution to encourage greater energy efficiency and use of least-cost strategies. The association also launched an investigation of the financing of Edison Electric Institute's lobbying campaigns. The investigation found that the majority of $26.3 million charged to consumers through their power companies in 1984 had been used to lobby for power industry positions—a direct violation of the rules for many states. The resolution to press for least-cost planning, if not accompanied by a cutback in industry influence, may turn out to be only good intentions. A report published by Critical Mass Energy Project in December 1985 found serious least-cost planning activity in only eight states—seven years after the passage of PURPA.

At bottom, consumers are not interested in participating in the affairs of regulatory agencies in an endless test of political influence. Their fundamental need is to use least-cost strategies to redefine electricity from a "commodity" to a "service." That may appear to be merely a boring distinction of economists, but these two opposing concepts are what the whole regulatory fight is about. In a Department of Energy poll taken in 1982, 60 percent of the consumers interviewed saw electricity not only as a service but as a "basic human right." This translates to service first, profits for the companies second. Private companies, however, need to have profits as their primary consideration; it's the reason for their existence and the focus of all their planning. For their purposes, electricity must be defined as a commodity tied to visions of expanding markets and increasing profit margins, not a service hinged on conservation, least-cost policies, and limiting the need for expansion—which limits the need for rate increases and profits.

It is at this fundamental level that the historical dimension of the industry re-emerges. Only in locally controlled, nonprofit public power systems and rural cooperatives is the concept of service the reason for the existence of a power system. While public and cooperative power systems may have strayed from this concept since they were "captured" in the 1960s, their economic base is still oriented toward electricity as a service and public rather than investor control. True to the predictions of John Bauer in 1950, the failure of regulation is giving way to increasing efforts for public ownership. Consumers in Long Island, New Orleans, San Francisco, Chicago, and dozens of smaller cities and towns are advocating buy-outs of private power companies and creation of new municipal or public utility district systems, which will be under direct voter control at the local level.

Opposing consumers in this effort to redefine electricity from a commodity to a service and to turn private power companies into public systems are not only the executives of the private companies but Wall Street investors and firms whose profits are tied to the power industry and its future growth. Greg Palast says, "It goes beyond the power of utility executives. In most states these guys are not very powerful figures. The powerful figures are the investment bankers behind them and corporations such as Bechtel and others who are making billions off these institutions." Pennsylvania Commissioner Michael Johnson speaks in a similar vein. He says, "Right now the utility industry has forgotten what its mission was. The mission was to perform certain kinds of services: electricity, gas, water. They now operate as handmaidens for the banking industry to provide them with lucrative investment fields. In the utility industry the chief thrust today is how many loans can be floated, how many bonds can be sold and at what kind of interest. Theirs is not a mission to try to get the lowest possible rate but the highest."

Similar to these allegations, some critics charge that investors who hold large blocks of power company stock still use the companies to milk consumers as in the 1920s. Others point to a pooling of interests and the historical momentum of the institutions that finance the industry as having a profound influence on the direction of the power companies. Behind the scenes, the workings of Wall Street are a prime factor in the power struggle taking place, and in the changes we may witness in the coming decade.

8. *Wall Street: The Stall of the Dividend Machines*

The system is so shot. There's no doubt in my mind that in five years, or certainly in ten years from now, none of us will recognize the structure of the utility industry in terms of anything we knew historically. But I can't tell you what it will look like. I don't know.

Eugene Meyer, vice-president
Kidder, Peabody and Co., Inc.

It was no coincidence that the Wall Street district of New York was the first area wired by Thomas Edison. Behind the publicity and false hoopla of the Pearl Street plant being the world's "first" power station stood J. P. Morgan as Edison's partner. Even before he had taken the train out to Edison's demonstrations in Menlo Park, New Jersey, Morgan had foreseen the profits to be made in creating manufacturing companies like Edison General Electric and, later, in financing and controlling electric power companies. In time, Morgan and other financiers came to realize that like the railroads, electric power companies provided an industrial heartblood that would be central to the nation's economy. Though the Morgan interests failed twice in attempts to take over the entire industry, his financial empire was to play a key role in its development, from the formation of the giant Electric Bond and Share holding company, which fought public ownership of power resources, to the rushed commercialization of nuclear power.

Twenty-three Wall Street, the site where Edison's lights first flickered on in J. P. Morgan's offices in September 1882, is now occupied by a broad rough marble building that houses Morgan Guaranty Trust, the nation's fifth largest commercial bank and an institution that holds control of more corporate stock than any other in the nation. Deep shadows fill the narrow street beside the low cornerstone-like structure, and the sky reflects off pale glass panels of buildings towering up thirty and forty or more stories around it. Directly across Broad Street is the New York Stock Exchange. And facing it on Wall Street is Federal Hall, where

George Washington was inaugurated as the nation's first president. An enormous amount of history has taken place at this corner, but only those familiar with the area would know the Morgan building. Here at the anonymous heart of the financial world little gold numerals on the door are its only marking.

In a larger sense most people have no idea of the role Wall Street firms and banks have played as silent partners in the power empire, the degree to which the money managers have shaped its course, and the stake they have in the industry's future. Simply put, power systems have been as much involved in the generation of capital as in the generation of electricity. And the course of the industry has been shaped as much by the need and interest of its main stockholders and financiers as by the demand created for electricity. Despite frequent denials, Wall Street has wrapped itself in the politics of the industry. In the struggle about to emerge over renewable energy and a rebirth of nuclear power, the role brokerage houses and financiers play will be critical. The events surrounding the private power companies, which make up the bulk of the industry, will be especially fateful.

For decades, private power companies have been known to insiders as the "dividend machines" of Wall Street—the most capital intensive businesses in the world—requiring $3 or $4 of investment for every dollar of revenue. In the process of building toward total assets valued at approximately $350 billion in 1985, the private power companies have absorbed a third of all industrial financing. Half of all new industrial common stock issued each year comes from private power companies. As much as half of all the investment banking business for firms such as Morgan Guaranty, Merrill Lynch, Salomon Brothers, Kidder Peabody, and other major Wall Street houses commonly comes from power companies. Financing charges for construction of a power plant can amount to 40 percent or more of its total cost. Fees are collected at every stage —for advising a company on its financial plan, for selling its stock and placing its debt, and from dividends on power company securities the brokerage house or bank might own. According to Kidder Peabody Vice-President Eugene Meyer, 40 percent of a consumer's electric bill is for financing charges and 40 percent is for fuel costs. In 1985 the 210 private power companies received nearly half of their $37 billion annual capital budget from outside financing. During the 1970s, as much as 75 percent of the industry's capital came from Wall Street and bankers. And during the past ten years, they have issued over $132 billion in stocks and bonds, more than any other industry in history.

This is substantial bread-and-butter money for the financial community. As a result, despite the passage of New Deal laws aimed at breaking the close relationship between the financial community and the private power companies, their ties have remained. Wall Street financiers continue to share a vision of prosperity dependent on increasing electric consumption. They fight to protect their current investments in the industry, and they guide the choices in financing that will help shape the industry's future. The heart of this relationship was molded by the social and political events surrounding the industry's earliest beginnings.

In 1873, just six years before Charles Brush's first power companies began operation, the New York Stock Exchange went through a total collapse. The financier Jay Cooke had failed with a massive Northern Pacific Railroad venture and took more than fifty companies down with him. The stock market closed for nearly two weeks to sort out the chaos, and a deep recession swept across the nation. As a result, investment capital in the 1880s was in short supply.

The usual practice in getting a private power company started was for representatives from Edison General Electric or Thomson-Houston or the Brush Company to organize a group of local businessmen, supply them with equipment and patent use agreements, and take one third of their common stock as payment. It was after more than 2,000 companies had been formed and twenty years after the lights came on in the Morgan building that ownership began to spread to a larger number of stockholders.

This change came suddenly as the "natural monopoly" concept took hold. Companies expanded, with long-distance transmission lines and new and larger generating stations. From 1902 to 1907, Morgan and others watched as the money invested in power plants increased from $500 million to $1 billion. Power output doubled. By 1912 both the value of the plants and the power output doubled again. At the end of the summer of 1913, Frank Vanderlip, president of the National City Bank of New York and an associate of the Rockefellers and Morgan, told a group of power industry officials gathered at Association Island, New York, that it was essential for the industry to attract more individual investors. Noting the heavy capital needs of the industry and its rapid expansion, Vanderlip said a crisis was approaching if new money could not be raised. Vanderlip estimated the doubling trend in the first decade of the century would continue:

> Four hundred million a year, eight million a week of fresh capital can be profitably used in the development of the whole broad field of the electrical

> industry in the United States during the next five years. . . . In making such an estimate, one does not need to draw on one's imagination. There is no need to picture broadening fields of application, new methods of production and distribution, nor new uses for current. A survey of what has been going on about us comes near enough to being a fairy tale. . . . Your dream is of great central stations that will radiate trunk lines; that will become as necessary in the lives of communities as are the lines of transportation; that will produce current which will become almost as essential to our every-day life as the blood in our arteries. . . .

Vanderlip pointed out that even though the railroads would use twice as much capital as the power companies in the coming decade, the power industry showed a far more rapid increase in earnings. The main obstacle he perceived was the corrupt image of the power companies and the competitive struggles over franchises. He said that the point had come where the very phrase "Public Service Corporation" carried to the investor's mind pictures of difficulties with boards of aldermen, threatened charters, expiring franchise rights, new forms of taxation, state-fixed rates, and profit divisions with municipalities before undreamed of.

But, he said, this image could be overcome by the establishment of new state commissions and the prospect of growth and profits to be made. Another part of the solution would lie in the growing political clout that came with influential investors. Vanderlip urged the utility executives to practice statesmanship as well as good business. If they acted wisely, he said, four out of five corporate investors could be persuaded to buy power industry securities and bring in the $400 million in fresh capital needed annually for the next five years.

Vanderlip's speech and others like it marked a turning point for the industry. Although stock market scandals in 1914, the struggles between private and public interests, and then World War I stalled the burst of investment, the emergence of the holding companies and the "golden years" of the 1920s would fulfill Vanderlip's prediction. Under "customer ownership" programs designed to build political support as well as funding for the private companies, large numbers of new investors would purchase stocks and bonds. But the concentration of control remained in a few hands through stock-watering schemes under the holding company structures. And the political clout of those holding control, such as Samuel Insull, increased enormously—establishing the private companies as monopolies and effectively stifling the growth of public power systems. With as much as one third of all the industrial financing during the 1920s going to power companies, the dividend machines were put together as devices that needed only a little central control and a generous amount of political grease. The usual mix of securities for private power companies became 25 percent in common stock, 15 percent in preferred stock, and 60 percent in bonds. This bond-to-stock ratio is heavier than most industries and reflects not only the long-term use and financing of the assets such as transmission lines and generating plants, but also the fact that smaller percentages of investors have voting privileges tied to common stock and can control a significant amount of capital.

Even after the collapse of Insull's holding company empire and the vicious legislative battles surrounding the passage of the Public Utility Holding Company (PUHC) Act of 1935, which forced the partial breakup of the conglomerates that controlled more than 80 percent of the nation's electricity, Wall Street firms and power companies continued to work closely together. The tenacity with which brokerage firms held to their ties with power companies was revealed in Congressional investigations that attempted for four decades to sketch an accurate picture of the institutional relationships and political influence held by the network of power companies and Wall Street firms.

In July 1935, as controversy over the PUHC Act mounted, the House Subcommittee on Rules probed the composition and support of the American Federation of Utility Investors, which was placing strong lobbying pressure on Congress to prevent the bill's passage. The federation was supposedly an independent organization with a national membership of 56,000. Under intense examination from subcommittee attorney William Collins, the president of the federation, Hugh Magill, reluctantly revealed that a number of insurance companies and an investment firm had made donations to the group, ranging between $500 and $1,000. As Collins pressed harder, Magill admitted that as many as a dozen brokerage firms had also contributed to the company. In explaining the source of other funds, Magill artlessly revealed that the organization was subsidized in large part by producing and selling pamphlets to power companies, which the companies then mailed to their investors. He also admitted that the organization's membership was acquired through the use of power company mailing lists. Although the group claimed to be an independent organization, it was clear that it operated in a close relationship with both the financial community and the power companies.

A million copies of one pamphlet, *An Appeal to Reason,* had been mailed out by the group just before a flood of letters and telegrams opposing the PUHC Act descended on Congress. Many of those letters and telegrams later proved to be fake. Magill also acknowledged that he had issued incendiary statements to the press against the legislation. One of his statements particularly outraged the subcommittee members: Magill had said, "Passage of this Bill means war. If war comes, investors will be mobilized and ready for it. The time of Congressional 'cracking down' on this great independent, unpampered class is past. Our turn to 'crack down' will come—if events force the issue."

The act, which was aimed as much at reforming Wall Street as it was at breaking up the huge combines of power companies, passed in the last days before the summer recess of 1935. In its final form, it did not contain a much debated "death sentence," which would have eliminated all utility holding companies. Instead, it broke up holding companies that did not operate contiguous power systems, reduced a dozen remaining major holding companies to a single-tier structure, and prohibited financial firms authorized to issue stock from owning voting shares of power companies. In setting up these legal barriers, the point was to limit the growing economic and political clout of the network of Wall Street agencies and power companies. Despite these measures the concentration of control remained through indirect channels.

A congressional investigation in 1940 found that brokerage firms continued as exclusive agents of the power companies. The firm of Morgan Stanley, for instance, which had previously controlled the United Corporation holding company, continued to handle all the bond issues for power companies once affiliated with the United Corporation. The president of Columbia Gas and Electric, one of the former United subsidiaries, admitted to a hearing panel that no firm but Morgan Stanley was considered to head Columbia's securities deals. Of a little more than $1 billion in electric company securities released between 1935 and 1940, Morgan Stanley handled $818 million.

Not surprisingly, the committee found that the ten largest insurance companies, controlled by the same financial trusts that had directed the holding companies, purchased 81 percent of the offerings. Morgan's Equitable Life Assurance Society, for example, was among the leaders in power company ownership.

The concern in Congress was that concentration of control by the financial community, although one step removed, remained intact. It resulted not only in concentrated economic and political influence but in inflated costs of noncompetitive underwriting and sweetheart stock deals to affiliate companies, which meant higher rates for electric consumers. Following the investigation, the Securities and Exchange Commission passed a rule to require competitive bidding in underwriting. But the problems continued.

In 1955, a Senate committee investigating the Dixon-Yates controversy found the two major holding companies operating in the South still had ties to financial institutions. The investigation had been initiated because of the political machinations of the private companies to stop the expansion of public power systems in the South. In probing the controlling interests of the Middle South and Southern Company holding companies (which before the PUHC Act of 1935 had been part of the giant Commonwealth and Southern combine controlled by the Morgan organization) the committee wrote in its report:

> Since passage of the Holding Company Act of 1935, the banks, investment trusts, and insurance companies have avoided individually holding enough voting stock [they were buying heavily in bonds and preferred stock] to be technically classed as a holding company or affiliate, which would require them to comply with the disclosure provisions of the act, as well as subject them to divestment of their ownership if the SEC found such action in the public interest.

But among affiliate companies, the committee found enough collective ownership of stock to control the power companies. The investigation also found "some of the same institutional investors own considerable blocks of stock in the two companies." Significantly, those investor institutions also had interlocking directors. The committee also discovered the two holding companies still shared a number of directors as well as mutual connections through regional banks. One witness in the case, J. D. Stietenroth, a secretary-treasurer for the Mississippi Power and Light Company, testified he didn't know when the companies had supposedly been broken apart because there was no change in the people he dealt with or in the transactions between the companies. He said that Electric Bond and Share Company (EBASCO), another former Morgan holding company, arrived at the power company offices to provide accounting and engineering services in the 1950s and the companies "just set out a barrel of cash, and EBASCO took what it wanted."

Suspecting that this was a situation not just peculiar to the South, the committee recommended a larger study to examine power company ownership throughout the country. The following year, the private power companies launched efforts to roll back the PUHC Act to allow the development of new private consortia to finance and develop nuclear power plants. The financial community, led by investment banker Lewis Strauss as head of the AEC, saw nuclear power as the capital intensive vehicle with which they would build a new privately controlled electric power empire. Strauss and former AEC chairman Gordon Dean, who now worked for Lehman Brothers brokerage house, assisted with the rewriting of the Atomic Energy Act to allow private development of nuclear power. Strauss also helped accelerate that development through support of the Price-Anderson Act, which provided the critical nuclear accident-insurance subsidy for the private companies.

During the Eisenhower years no further study of the relationships between the financial community and the power industry was undertaken. But in 1965, as new holding companies began to form with the new power pools, there came a recommendation from the House Judiciary Committee for investigations. In 1968 the SEC conducted an initial study and reported in 1971, citing serious "gaps in information." The committee requested further detail from the companies on their stock ownership and relationship to banks and brokerage houses. A Senate committee took the new information in 1974 and held Corporate Disclosure hearings to investigate the concentration of control in a number of industries including electric utilities.

Congressman Michael Harrington of Massachusetts, testifying in the Corporate Disclosure hearings on the New England Electric System holding company, said the company had direct and indirect interlocks with thirty-seven banks, insurance companies, and law firms, and thirty-one interlocks with other utility companies.

In a letter to John Nassikas, head of the Federal Power Commission in 1974 (who would be appointed as a special commissioner in the Seabrook case ten years later), Harrington pointed out:

> The dangers inherent in interlocks between banks and public utilities is equally serious, if not more serious, today than they were in the 1930s. Regulatory practices adopted by the Federal Power Commission and most other state public utility commissions permit utilities to base their ratio of rate of return on the cost of their capital. In other words, the higher the interest rate, the higher the profit rate. By permitting banks and utilities to be controlled by the same directors, the result will naturally tend toward higher rates, and consequently, higher utility profit margins and bills.

Harrington and others at the hearing pointed out that the interlocking power companies were connected not only to the same banks and brokerage firms but also to the same fuel companies and engineering services. The sharing of business information in such relationships, they believed, had a great danger of stifling competition, developing monolithic policies in the industry, inflating prices, and ultimately creating even greater centralization. The interlocking power companies could also wield substantial political clout in supporting election of state and congressional officials who would support and not question their policies.

In terms of the larger political effects, Harrington lamented that there had been a steady momentum to undermine the spirit of the PUHC Act of 1935. Between 1936 and 1966, the SEC had amended its rules ten times to provide greater exemptions from the prohibitions on interlocks. And in violation of the intent of the PUHC Act and antitrust laws, it was estimated that as many as 1,000 bankers sat on the boards of the power companies.

A subsequent series of investigations conducted in 1978 and 1979 by the Senate Committee on Government Affairs detailed the interlocking directorates and stock concentration further. In a report the committee published in 1980, the J. P. Morgan organization was shown to be still the number one stockholder in General Electric, number two in its chief competitor Westinghouse, and by far the largest stockmarket investor in the nation. Interestingly enough, GE and Westinghouse shared five of their top fifteen investors in 1980—a fact that underscores the monolithic policies of the industry. The Senate investigations also showed that Morgan managed portfolios in twenty-six utility companies. Directors of associated companies also sat on the policy-making boards of thirty-five major power companies.

The J. P. Morgan Company's involvement with the electric industry, although singularly significant, is only the beginning of the financial community's continued ties to power companies.

Insurance companies had more than $90 billion invested in electric utilities in 1985, with Metropolitan Life and Prudential Life among the leaders. Many brokerage firms and banks that were making bread-and-butter money on power companies also held stock in the companies under "street names" that veil the actual owners.

The influence of the power industry and Wall Street is perhaps best reflected in the fact that after four decades of hearings and probes, the federal government could produce only a surface sketch of who controlled these major industrial corporations and the nature of the interlocking relationships between the corporations. Members of the Senate committee shared a common fear over the deepening political and economic influence of these corporations and the individuals who controlled them. In the 1980 congressional report, they urged initiation of a more in-depth study similar to that of Roosevelt's Temporary National Economic Committee. Members wanted to create a more thorough outline of the stock ownership, debt ownership, interlocking directorates, and other business relationships. It was hoped that the information would be fed into a computer, which would help provide a detailed picture of the control of the corporations and their impact on the economy.

Following the election of Ronald Reagan, however, not only was further study pushed aside, but in April 1982 John S. R. Shad, an E. F. Hutton executive appointed by President Reagan to head the SEC, accepted the urging of his brothers on Wall Street and the power industry and recommended that the PUHC Act of 1935 be repealed. Similar proposals emerged in 1983 and 1985 and were stopped by coalitions of consumer and environmental groups. The fact that the top official of the agency responsible for maintaining this fundamental barrier advocated its removal marked a significant turn-around from the investigations and warnings of the 1970s. It also indicated the persistent intent of the power industry and financial community to work more closely in the future and raised a warning sign of a larger clash brewing.

□ □ □

The electric power industry is no longer run by a few domineering men in brown derby hats at the corner of Wall and Broad streets, but the mutual interests to keep the dividend machines running remain. The power companies usually take their financial plans to their investment bankers in the fall of each year. In advising company executives on the timing and arrangements for their financing, the investment bankers claim they have no role in the projects themselves. Brokers who sell the securities and analysts who publish reports for investors claim they have a sacrosanct role just to "provide money" for clients with viable projects. But in everyday business they function as promoters and defenders of power company projects and as key players in the future directions of the industry. As growth of the power industry has stalled and the crisis surrounding nuclear construction has deepened toward a $100 billion to $200 billion damage bill, the workings of Wall Street at the local and regional level to protect their interests has become startlingly visible.

The foremost example of Wall Street's actions came in the financial collapse of the Washington Public Power Supply System in the Northwest. In 1981, two years before the landmark $2.25 billion default, citizens attempting to stop runaway spending on construction of the five reactors mounted a statewide initiative to require voter approval for any additional bonds to be issued for the project. Nuclear contractors working on the five plants responded with a countercampaign. Prominent Wall Street firms, some of which claimed to make the highest profits in their history selling the steady stream of WPPSS bonds, contributed heavily to the countercampaign. According to the state's Public Disclosure Commission, Merrill Lynch and Pierce, Fenner and Smith, Inc., donated $20,000 to stop the citizen initiative; Smith Barney, Harris Upham and Co. $15,253; Goldman, Sachs and Co. $15,000; Salomon Brothers $15,000; Kidder, Peabody and Co. $10,000; and Paine Webber $10,000.

While citizens won the initiative, the countercampaign was later successful in overturning the new law in federal court and in having that decision upheld by the Supreme Court. Nearly $1 billion in additional bonds was issued before a power surplus (triple the needed reserve), a consumer rebellion, and state court decisions brought the project tumbling down.

In October 1983, three months after the massive default had hit Wall Street, Chemical Bank, the bondholders' trustee, called a national meeting in New York. It was held in Madison Square Garden's Felt Forum and surrounded by all the technological pomp and ceremony of the media age. On stage Chemical Bank Vice President William Berls appeared on two ten-foot-high color television screens and pledged to "take any actions necessary" to force consumers to pay for the cancelled plants. Berls and four other Chemical Bank officials were being broadcast live from a nearby television studio and transmitted via satellite to meeting rooms in Chicago, Los Angeles, and Seattle.

For well over an hour 1,200 analysts, observers, and bondholders sat in the dim and airy 4,200 seat auditorium listening as the panelists provided a summary of the WPPSS project from its inception in 1968 to the financial collapse in 1983. For another hour the officials answered live questions, phoned in from the meeting rooms. Ralph McAfee, an attorney for the company and one of the panelists, told viewers he would like to see consumers in the Northwest "punished" for their refusal to pay for the cancelled plants. Berls called the default a "national crisis" and laid out a plan for a national network of WPPSS bondholders, estimated to number 75,000, to organize and apply pressure to Congress for passage of bailout legislation. Aside from helping write the bailout legislation, Chemical Bank also mounted a court suit charging 600 public officials with fraud which was expected to take until 1988 to litigate.

Chemical Bank's WPPSS meeting was a televised sample of what was going on in board rooms all across the country as billion dollar nuclear plants were cancelled and institutional investors and Wall Street firms attempted to prevent massive losses and a shift away from nuclear technology. Although it was a highly visible spectacle, what unfolded at the

Felt Forum was not an isolated attempt to pressure consumers into paying for the industry's overbuilding and overfinancing.

In the spring of 1984, on the opposite coast, Merrill Lynch Capital Markets, the investment banking division of Merrill Lynch, created a plan to save the Seabrook nuclear project and its investors from financial catastrophe. The three-phase plan, from which Merrill Lynch would make $100 million, called for the creation of a new corporate entity "Newbrook," extension of over $200 million in loans, and a billion dollars in additional financing. While the plan would save the investments of stockholders, commercial banks, Prudential Life's Prulease Corporation, and European lenders, it would do so at the expense of consumers and taxpayers. The plan also added to the cost of electricity from the plant (estimated at between four and eight times that of oil-generated electricity), drained off capital needed for conservation, and enlarged the power surplus in the region.

Other faltering nuclear plants such as Long Island Lighting Company's $4.5 billion Shoreham project brought similar bailout proposals. The forerunner of these financial rescues had been hammered out in the aftermath of the 1979 Three Mile Island accident. In order to save the plant's owner, General Public Utilities, from bankruptcy, a consortium of major banks provided loans to halt the company's financial plummet. A broader strategy of the rescue plan was aimed at stopping the flight of investors from other power companies building nuclear plants, as well the sinking of stock prices for uranium suppliers and nuclear equipment manufacturers like GE and Westinghouse. In hindsight, this action to save investors may have forestalled natural market forces; if allowed to continue, a slide in stock values could have slowed the continued spending for nuclear construction and the loss of untold billions in additional debt—all for plants that now stand abandoned or canceled because of safety questions or a lack of need for their power.

By the spring of 1984, trouble over safety issues and soaring costs threw the completion of more than half the forty-eight nuclear plants under construction into question. Given the serious problems facing plants like Seabrook and Shoreham in the Northeast, Grand Gulf in the South, and Palo Verde and Comanche Peak in the Southwest, proposals for a national bailout of the industry began to emerge.

In the fall of 1984 Eugene Meyer, a vice-president and investment banker for Kidder Peabody, began organizing a group known as For Responsible Energy Action. Its purpose was to "educate" Congress on what many Wall Street executives claimed as a dire need to save troubled nuclear construction projects. Meyer, a late-fortyish, balding executive with large metal-rimmed glasses and a good-humored manner, is the director of Kidder Peabody's Utility Corporate Finance department, which is engaged in raising new capital for power companies. Meyer is considered to be one of Wall Street's most vocal commentators on the power industry.

In Meyer's view and that of most power industry and Wall Street executives, zealous regulation at the state level—and not overbuilding and overfinancing or soaring electric rates—is to blame for undermining investors' trust and bringing about the collapse of nuclear construction. The tremors that had begun with WPPSS had within two years ended up in the heart of the nation—Indiana. "The last vestige of trust was blown out of the water in Indiana," Meyer said. "Their reputation, from an investors' point of view, was of having the best regulation in the nation—bad things don't happen to you in Indiana. Those are good, solid people. Good solid business folk."

In Indiana, Republican Governor Robert Orr had appointed a panel of five business executives to study the issue of what to do with the 59-percent-complete Marble Hill nuclear plant. Public Service Company of Indiana wanted to double rates in order to finish the plant. The governor's panel found that the power from the plants was not needed, and fearing dramatic rate increases, industrial plant closings, and economic recession, they recommended canceling both units and settling the losses on investors rather than consumers. Marble Hill was the beginning of a final wave of trouble that hit partially completed plants in the North Central states and raised the ante for Wall Street to continue financing the projects.

Meyer believes that, after the shock of WPPSS and Marble Hill, it will take two major steps to restore the system. The first is a bailout of troubled nuclear projects. "What we want is to create an awareness at the national level of the severe economic dislocations that would result from the problem of unfinished nuclear plants," he says. From his view "tremendous" tax, job, and private financing losses will ripple through the economy if the "roughly $80 billion to $100 billion in unfinished plants" has to be absorbed by the United States economy.

A second step calls for setting up a "blue ribbon commission" or at least getting assurances from the White House, Congress, and the Supreme Court that state governments will not be able to stop companies from charging automatic rate increases for plants being constructed. In addition Meyer also wants tax credits and greater investment incentives to help fund new building programs.

In the long term, Meyer maintains, the future expansion of American industry and the economy are at stake. In his view people are "misled" on the potential for conservation. "Electricity is a central thing to this country," he says. "It is not adequate to say 'Oh, conserve a little more and it will be all right.' No, you've rent the fabric, and you damn well better sew it up."

The campaigns against consumers, the bailout proposals, and the "educational" effort spearheaded by Meyer are only a few of the more visible occasions on which Wall Street firms have intervened in choices being made over the nation's future energy supply. Generally the workings of Wall Street agencies are not quite as dramatic. Vice-presidents and analysts from Kidder Peabody or Morgan Stanley or other major firms usually work with quiet precision, appearing in state or federal hearings to provide financial rationale in support of the lobbying efforts of the power companies for higher rates or new plants. One insider, disgusted by Wall Street's role, calls their work "numbers pasted over ideology." Another who has worked as a rating agency analyst says the lack of objective analysis of power industry projects and financing is "incredible."

In addition to a presence before state regulatory commissions, Wall Street firms are also active in electoral politics. Senator James McClure of Idaho, head of the Energy and Natural Resources Committee, who had received $46,775 during the 1981–82 and 1983–84 election cycles from utility companies and nuclear firms, also received $34,275 in 1983–84 from Wall Street firms, banks, and major insurance companies. Senator John Tower of Texas, who also serves on that committee, received $44,800 from the same financial firms. When the Long Island Lighting Company was facing bankruptcy over the $4.5 billion Shoreham nuclear plant, it's no secret that company officials and influential New York financiers turned to Senator Alfonse D'Amato of New York, who introduced and won a bill to allow special tax-free financing for Shoreham. D'Amato received $65,400 from the financial community in 1983–84. D'Amato also introduced bills to repeal and amend the PUHC Act in 1982 and 1983. McClure introduced a similar bill in 1985.

In the guts of it, the influence of Wall Street has historically translated into a bias in favor of high-cost, centralized power alternatives such as nuclear plants, as opposed to decentralized solar and conservation. This has been evident since Chase Manhattan's excitement over nuclear power in the 1950s. At the local level it means that banks continue to encourage the construction of inefficient all-electric homes, which make up more than half of all the new homes being built in the 1980s. In the larger political realm, it means that a company such as Merrill Lynch uses its extensive contacts and influence to work for a billion dollar bailout of the Seabrook project. It's also visible in Wall Street's lobbying to pass laws to speed reactor licensing, gut the Holding Company Act, reduce safety regulations, or provide greater tax breaks for power company investors —in order to keep the dividend machines running and start the flow of bread-and-butter money for financing again.

On the other side of dividend machines, there is a different reality. Consumers seek minimized costs, fewer environmental risks, and adequate public control over a service that has become central to their lives. Those who have been active on the issue believe that the crash of the nuclear industry is the result of too little regulation and too much self-interest from the financial community. Many support the claims of energy experts like Amory Lovins, and those of the Congressional Research Service, that with adequate conservation programs and utilization of waste energy, no new generating plants will be needed until after the year 2000. In their view, Wall Street executives like Eugene Meyer fail to recognize that, if investors don't eat the $100 billion to $200 billion damage bill, consumers will have to pay soaring prices for power—posing equally disastrous impacts on electricity-using industries and people who simply can't afford higher bills.

As they watch the growing consumer unrest and the efforts for greater public control, some individuals on Wall Street believe that a change in attitude may be just around the corner.

□ □ □

Two blocks from the site where Edison's lights first flickered on in Morgan's offices, Eileen Austen, a manager of municipal bond research and vice-president of Drexel Burnham Lambert, Inc., occupies a cubbyhole office off to one side of a large room filled with long desks and banks of computer terminals. The room is awash with the chattering of salespeople buying and trading over telephones. In her mid-thirties, Austen is very poised, articulate, and careful. As a municipal bond specialist, she wrote hard-hitting reports to investors on the approach of the Washington Public Power Supply System default and later testified before Congress against a federal bailout for the project. Her analysis of the power industry's crisis presents a stark contrast to Eugene Meyer's.

In Austen's view, the WPPSS project's financial collapse was brought on by the failure of the institutions to respond to a changing environment. Private power companies, encouraged by tax and accounting incentives, and public systems accepting high growth forecasts were overly eager to expand. The state regulatory bureaucracy and the federal government didn't provide adequate oversight. And Wall Street, suffering from a kind of institutionalized lethargy and interest in supporting the status quo, let things fall through the cracks. "The situation was one in which people were unwilling to face up to the fact that there were major problems," she says. One shortcoming amplified the next and ultimately criticism and decisions to halt financing came four or five years too late when debts on particular projects had grown enormously bloated.

Similar to this particular failure of financial and regulatory institutions, other weaknesses within Wall Street contributed to the national crisis. The predominant ideology that favored large capital-intensive projects was one of the deep internal problems. Some analysts contend that this attitude translated into serious contradictions for day-to-day business. Managers of brokerage firms were sometimes faced with the conflict of having salespeople promote profitable sale of a stock or bond in one department while analysts in another department were compiling a report that exposed critical weaknesses in the same offering. The pressure to sell often took priority.

In some cases, analysts were fired or pressured to quit because of a refusal to soften their analyses to please sales departments. Ridicule by other analysts also occurred. On one occasion, at a luncheon for a group of security analysts, Austen says the chicken entree was named in her honor and the veal for Howard Sitzer, another analyst critical of power projects top-heavy with debt. Such petty events begin to show the depth of the prevailing attitudes and the social pressure to maintain the status quo.

Another professional analyst, who requested to remain anonymous, had previously worked for Moody's Investor Services. He noted that equally serious conflicts exist within the rating agencies. The industry's two main rating agencies, Standard and Poors and Moody's Investor Services, are private businesses paid by issuers to provide a grade on new securities. The securities are generally rated on a letter system from A to D, with AAA being the highest recommendation for purchase and anything below BBB considered speculative. The ratings are used by investors as a primary recommendation for or against an investment. According to this analyst, the rating agencies are reluctant to provide less than an A rating on bonds, because that would be an insult to the issuer.

When this analyst was first hired by Moody's he said he had never seen a bond. A month later he issued his first report—a critical one. The bond somehow still received an A rating. Austen said she and her colleagues read that particular report and were mystified by the unwarranted rating. Under such a system, billions of dollars in power company financing can be issued and purchased by unwitting investors every year.

Jeffrey Whitehorn, an analyst with Dreyfus Corporation, confirms that the marketing of bonds in some cases is "completely incompetent." Whitehorn recalls that at one point in the WPPSS bonding, the financial people from WPPSS were appearing on the market every forty-five days or so with more bonds. "They would have their 'dog and pony' show, their bad food and bad cigars, and give the same old lines. It was a billion dollar project being managed by farmers and muffler salesmen who had no idea of what they were doing." Whitehorn says the problem with the rating agencies, along with hiring inexperienced staff, is that they are not anticipatory and only respond after disaster has struck. Six months before the catastrophic default and after two plants had already been canceled, WPPSS bonds were still enjoying an AA rating from Standard and Poors. A lot of small investors who only looked at the ratings, unaware of the inside news, got chewed up and spat out.

The economic collapse of the WPPSS project, the subsequent rage of both investors and consumers, and a federal investigation has provided motivation for tighter analysis. For most analysts it has forced a separation of power companies into categories of those whose survival is at stake, those who face a decade of low earnings, and those positioned for growth. But the overall picture is one of uncertainty over what the future holds.

In Austen's view the crisis over abandoned and canceled nuclear plants has created a permanent shift in investor support for power companies with nuclear projects. She believes that power companies will make a technological transition to use more solar and renewable energy. Charles Silberstein, an analyst for the Finance Guaranty Insurance Company, agrees. He says the highly visible and agonizing collapse of WPPSS may have marked a turning point and stopped other power companies from cutting their own financial throats. But there are other observers who say that in many ways, nothing on Wall Street has changed. Howard Gluckman, a writer for *The Bondbuyer,* observed that there are investment bankers who looked at the fiasco and said, "Nothing went wrong. We all made money didn't we?"

S. Arlene Barnes, an analyst for First Boston Research Corporation, is among those who believes the change is temporary. "I don't believe that anything is permanent," Barnes says; "the market will be selective but confidence will spring back." Mark Luftig, a vice-president and analyst for Salomon Brothers, agrees with Barnes's view. Luftig says that in the wake of the Three Mile Island accident, the market bounced back in about eight months. "It happened in March and by November or December you couldn't tell the difference," he says. "That was a problem with a single plant. The situation we're looking at now is with a number of plants. It's going to continue for the next two or three years and then move ahead."

In order for investor confidence to return by the late 1980s, Luftig says several changes in government policy will need to be in place. Echoing Eugene Meyer, he says licensing of nuclear projects will have to be streamlined, and regulations for bringing a plant on line will have to be expedited.

Austen agrees that the regulations and policies of the federal government are critical, but she sees rising costs and consumer anger as the deciding forces. If the government were to provide incentives for conservation and renewable energy, Austen believes a transition would be quicker. However, even if those incentives aren't put in place, she says, the handwriting is on the wall. "It will happen sooner, or it will happen later," she says. "There will be a learning curve as in any other industry."

How slow that learning curve is may determine how sharply rates will increase for consumers. Austen says the power companies that don't push ahead with conservation and renewable energy programs could lose the confidence of their investors and end up paying more for capital from high-rolling risk takers. The higher cost capital and a need for new plants will in turn raise rates. At this point consumers will provide the company's incentive to change. "People will resist paying for dramatically increased rates for power," Austen says. "It's what we're seeing all across the country. The industry is being forced to scale back and will continue to scale back."

In cash projections compiled by *Electrical World* magazine in September 1985, scaling back from the overbuilding of the 1970s (in 1985 dollars) is already extensive—from $49 billion in construction in 1982 to a low of $17 billion in 1991, then arcing upward again to $48 billion by 1999. The dividends will keep flowing, but the growth of the dividend machines will be stalled. In view of these spending cuts, Wall Street executives like Meyer and Arlene Barnes expect to see rate reductions, rather than rate increases, becoming a trend around 1987. They hope that a drop in rates will stimulate demand and spur the need for more plants.

In the meantime, those who support greater centralized growth are lobbying to eliminate government regulations that allowed public intervention and slowed the construction of plants in the 1970s and early 1980s. Meyer's campaign is a visible part of this. Others, such as John Huneke of Morgan Stanley, Charles Benore of Paine Webber Mitchell Hutchins, and Robert Burke of Moody's Investor Services are also pressing Congress to streamline the process for reactor licensing, to provide tax credits and construction-work-in-progress (CWIP) charges, and for other changes that will help restart the building programs. But the options for new cogeneration and energy efficiency, coupled with public anger, may be undermining their influence.

Among alternative energy advocates, there is talk that Wall Street may not be needed to play a primary role in a technological transition to solar and renewable energy. Amory Lovins envisions a financing system in which power companies could offer loans for consumers to weatherize or solarize their homes instead of building new power plants. In his estimate, much of the capital for a "revolving loan fund" could come from the company's internal sources. For those power companies that have been drained of cash by ill-fated nuclear or coal plants, he believes initial capital could come from ten- or twenty-year public sector loans similar to $500 million bond issues Oregon and California are already using to finance energy investments.

According to his calculations, undertaking a broad energy-efficiency program could cut environmental and social costs of acid rain and radioactive wastes, and allow consumers to save more than $10 billion per year on their electric bills. Additionally, the trillion dollars in capital which would have been drawn primarily from Wall Street for expanding the power industry under Department of Energy recommendations for some 200 new coal and nuclear plants would be available for other investments.

The motivating force for the revolving loan funds would be contained in orders from state regulatory commissions that power companies pursue least-cost planning. This would ensure that conservation and renewable energy would be considered before construction of new power plants. Lovins believes that the financial community will accept such measures because they would stabilize precarious utility financing and deter power companies from construction programs they cannot afford. Other alternative energy advocates are equally optimistic, citing utilities in California, particularly Southern California Edison, as setting the trend for the future. They also point to alternative-energy funds being established by Merrill Lynch and Kidder Peabody. But some insiders have been much more skeptical.

John Dorfman, a Wall Street analyst and author of *The Stock Market Directory,* wrote in 1982:

> Most of what you'll read in the papers about Southern California Edison concerns the company's commitment to alternative energy sources like solar energy, wind and biomass. . . . This goes over big, especially in California, but it ought to be put in perspective. In the early 1980s, the company had the capacity to generate more than 13,000 megawatts of electricity daily. It expected to add about 5,500 megawatts of capacity in the 1980s. That would make a total of around 18,500 megawatts of which only 2,000 or so would be from the new style sources. . . . The pledge to emphasize renewable power sources in the future may be of value in getting potential help (or at least leniency) from state regulators whose aid the company (like almost all utilities) desperately needs.

Hopes may have grown brighter since 1982, and the developments in California may prove Dorfman wrong, but for most states there is serious uncertainty as to whether regulatory agencies will indeed force power companies to pursue least-cost options with vigor. In all likelihood, cogeneration, solar, and efficiency improvements will take place, and power companies will see 10 or 20 percent of their growth met by these sources. But the bias toward large capital-intensive centralized projects and the plan to revive building programs may distort these efforts and other renewable energy projects, making the least-cost programs little more than public relations gimmicks. Unless regulatory agencies force least-cost alternatives or there is a shift toward more direct public control, it's doubtful that these sources will be utilized extensively enough to break down monopoly control and act as a technological bridge for a renewable energy transition.

Further, while a few Wall Street executives boast about the amount of financing they are making available to solar projects or cogeneration, it's clear that they are not mounting "educational" campaigns to support the need for solar technology. In the fall of 1985, when Congress continued tax subsidies for nuclear, coal, and oil plants and scheduled elimination of subsidy for solar, conservation, and other alternative power sources, Wall Street did not take a position, even though many small firms were being forced out of business. It was clear that alternative energy development in the United States was being slowed down. The choice for what technological options and power sources would be available in the late

1980s and early 1990s was being made with renewable energy left behind.

□ □ □

The dream of the "golden days" of the industry is what still possesses power company officials and their friends on Wall Street. In the 1920s and again in the 1950s and early 1960s, it meant steady growth, comparatively little government oversight, and stable fuel supplies. Despite the fact that the power industry has matured and reached a level of slower growth, industry officials hope to re-create those eras with the "second coming" of nuclear power in the 1990s. Encouraged by the deregulation policies of the Reagan Administration, the power industry is preparing for the renewed construction program in the same way it did in the 1920s —by expanding its political and economic base.

In addition to the industry-wide attempts to roll back the PUHC Act, there has been a return of customer ownership programs, re-emergence of stockholder organizations and other pro-utility groups such as Eugene Meyer's For Responsible Energy Action, and requests to expand loopholes for interlocking directors. In June of 1986 GE purchased RCA Corp. and its NBC television and radio networks. And in the most significant change between 1982 and 1986, there were some thirty new electric holding companies formed. Added to the nine existing major electric holding companies (Southern Company, Middle South Utilities, Northeast Utilities, Eastern Utilities Associates, New England Electric System, Central and Southwest, Allegheny Power, American Electric Power and General Public Utilities—which control some sixty utility companies) are now some 110 smaller holding companies—more than half the industry.

"It's beginning to look like a return to the old days," one federal official observes. But more than a return to the old days, the activity signals a new era of corporate centralization about to sweep through the industry. More than just a choice between solar or nuclear technologies, the changes now stirring contain substantial meaning for the question of who will control the electric industry and an increasingly central part of the nation's economy.

Industry officials and Wall Street executives claim the companies need flexibility to install more directors from the financial community, create new financing mechanisms, and get involved in other businesses without interference of government oversight. In states where they have formed new holding companies to launch such efforts, the initial response from consumers and business organizations was strong opposition.

In 1982, Central Maine Power Company proposed a holding company to be known as Maine Industries, Inc. The businesses to be included under the new corporate umbrella were a real estate venture, timber harvesting operations, a finance company, and hydro and cogeneration projects. Consumers and local business organizations bought full-page newspaper advertisements criticizing the plan. Businesses objected that the power company could use its financial weight to unfair advantage. Consumers feared that, if the other ventures were successful, they would not benefit from the profits, whereas if the ventures failed, losses would end up in their bills.

Maine, like many other states, had no authority over such plans, but the state passed emergency legislation, giving oversight to the Public Utilities Commission. The commission turned down the plan.

Similarly, Public Service Company of New Mexico proposed diversifying its operations to include a real estate firm and a fiberboard manufacturing plant. Farmers in the state were worried by the company's efforts to purchase water rights in key areas. Business people were also concerned. A fifteen-month moratorium on power company diversification was declared while restrictions were put in place.

In New York, the Long Island Lighting Company sought to start a solar business and mailed out flyers to all its customers. Local solar companies claimed unfair competition and charged that the power company didn't want to supply conservation or solar equipment to homes but to create a stall in the market, undermining their businesses and the trend to decreased use of electricity. Enough questions were raised that the New York Public Utilities Commission eventually ordered the power company out of the solar business.

These examples are the proverbial tip of the iceberg. In nearly every state similar expansion efforts, outside or inside the power industry, are going on. Guy Nichols, former head of the New England Electric System, told a congressional panel in October 1983 that 63 percent of the power industry is now diversified. Many of the holdings are in fuel, mining, and engineering companies. Other prospective ventures are as diverse as owning a baseball team (a proposal of Florida Progress Corporation, the new holding company for Florida Power and Light) and ventures into telecommunications.

As the stall of the industry has deepened, regulators and alternative energy advocates have slackened their opposition to diversification, mergers, and the creation of new holding companies. In Florida, as in a number of other states, objections to holding companies have faded as alternative energy advocates have been persuaded that the private power companies will use the opportunity to launch serious alternative energy subsidiaries. In the summer of 1985, however, the Florida Public Service Commission announced an investigation into indications that the three electric holding companies formed in the state since 1981 were using their conglomerate structure to pass overcharges from fuel subsidiaries to consumers and make high profits.

At the same time in Ohio, strong controversy broke out over a proposed merger for two former General Electric subsidiaries, Toledo Edison and Cleveland Electric Illuminating Company. The new holding company, which might eventually include Ohio Edison, was being formed to bail the companies out of trouble in financing the completion of the Beaver Valley and Perry nuclear plants.

The company promised $90 million in savings for consumers. The state consumer counsel at first opposed the plan but, reportedly under intense pressure, turned around to support the deal. Many consumer advocates remained skeptical. The new generating subsidiary to run the nuclear plants would be under federal rather than state regulatory control, guaranteeing weaker regulation and higher rates. Further, public power systems such as the Cleveland municipal would be at a disadvantage in purchasing power from the private companies at competitive rates now that they operated under a single roof. The merger also gathered considerable political influence, which some critics said explained why Governor Richard Celeste and the state consumer counsel backed the deal. In the long term, the merger will affect the politics of the region's power pool and an agreement with Ohio Edison, Pennsylvania Power Company, and Duquesne Power and Light for sharing of power and transmission lines. Speculation has emerged that Ohio Edison may join the new holding company in the not-too-distant future.

Despite the prospects for alternative energy development and the promises of cheaper rates, creation of new holding companies holds long-term danger for consumers. One primary interest in forming new holding companies is the ability to place generating plants under a single subsidiary and escape state regulation. Under this scheme, the generating company sells power under wholesale contracts to its sister distribution subsidiaries. Wholesale power rates come under the jurisdiction of the Federal Energy Regulatory Commission (FERC). In addition to allowing higher rate increases, FERC also allows 50 percent construction-work-in-progress charges—allowing the new holding company to bill for cost overruns that might be refused by a state commission.

Creation of new holding companies and diversification into other businesses begins a shift of regulation away from the state agencies to the federal level, thus eroding the possibility for public control. This is the first step toward a larger change Wall Street and the power industry would like to see—one that would finally produce the political overhaul of the dividend machines they have sought for decades.

The events now unfolding are being shaped to fit neatly into the industry's longer-range plans to win price deregulation and a restructure of the entire electric industry into generation companies, transmission companies, and distribution companies. Under such a scenario, insurance companies, major oil companies, banks, and brokerage houses could take part in conglomerates that would construct four or eight nuclear or coal plants in "energy parks"—regional sites ultimately managed by only 15 or 16 mega-holding companies—fulfilling the dream that floated through the industry and the halls of the federal government in the late 1960s and early 1970s.

The only regulated portion of the business would be the local distribution companies. In theory, prices would be kept low by competition between the generating companies, but there may be little leverage to prevent the rise of an electric cartel. While such a restructuring is impossible at this stage, the federal government is allowing a few experiments to test the viability of deregulation, which the Reagan Administration has promised to bring about.

In order to build support for their plans and to help neutralize opposition, power companies have boosted their public relations efforts and reintroduced customer-ownership programs. By the end of 1985 more than a dozen major power companies were instituting programs in California, Montana, Washington, Idaho, North and South Carolina, Virginia, and Maine.

An official of the Virginia Electric Power Company said, "It's a good public relations tool. It's a way of bringing customers over to Vepco's side because they begin to see why we need a return on equity." Between mid-1980 and mid-1983, Vepco (which had reorganized into Dominion Resources holding company) reported raising $24 million in the sale of 420,000 shares of stock to 33,200 customers.

Dennis Lawler, director of the financial services department of Wisconsin Electric Power Company, said his company's program, like others, sends out literature on the program with customer bills. He said the program in Wisconsin was not based on the need to raise capital but on "getting customers to have a stake in the company."

The San Diego Gas and Electric Company program is also similar to other customer-ownership programs. Lynn Taylor, the company's financial communications administrator, said, "We are trying both to raise revenue and get customers to have an interest in the financial well-being of the company. It will also help neutralize the problem of our customers only reading criticism in the press every time we file for a rate hike."

In addition to "neutralizing" opposition, these local stockholders can be further organized into regional or national stockholder groups to apply pressure to Congress. In April 1984, James Spang, head of an organization known as the American Society of Utility Investors, appeared before a Senate committee to argue in favor of federal institution of construction-work-in-progress charges. Spang, who holds a doctorate in education and works for the Pennsylvania Education Department, describes himself as "pro-business and pro-American." His organization, which claims 5,000 members, was chartered in the wake of the Three Mile Island accident in order to press for rate increases to keep stockholder funds flowing from General Public Utilities Company. Spang, like investment banker Eugene Meyer, believes that government assurance of profits for power companies and their investors is essential to making the nation's power systems work. He told the Senate Subcommittee on Energy Regulation that because of tightening cash flows, inflation, environmental legislation, and safety considerations, "Utilities had no choice but to request rate relief." Following their urgent requests in the 1970s he said, "New notions about the nature of democracy led to rate resistance and other forms of consumer activism which, in turn, led to further rate relief requests, abandoned plants and scared investors." Spang makes no mention of the surplus in power capacity that approached 40 percent in 1982 and the fact that the utilities were overfinanced and overbuilt.

In addition to appearing before Congress, Spang claims that his group has been instrumental in adding to the rate increases allowed by state public utility commissions. At an annual meeting of General Public Utilities, Spang credited his group with adding an extra $7 million to rate increases in a New Jersey case.

While local or national investor groups may have political clout, they have little control over a power company's policies or operations. The bulk of a company's stock is still held in a few hands. According to Edison Electric Institute there are some 8.8 million individual stockholders in electric power companies. Other estimates, which include investors in funds managed by trusts and mutual groups, range to 35 million. Most of these people, however, have no voting rights or voice in power company policies. At the top, the stockholders electing the company's board of directors are representatives of large institutional groups or managers of individually held accounts. These top stockholders are often anonymous, listed in a "street name" of an undisclosed individual or individuals whose account is managed by a brokerage firm or bank. Although the top stockholders usually control the company with combined holdings of less than 10 percent, ownership and control can be much more concentrated. The top nine investors for Commonwealth Edison Company of Chicago in 1982, for example, controlled 47 percent of the company's voting stock. Among them were street names such as "Calder & Co." listed under the Bank of Nova Scotia at 67 Wall Street in New York. Other top stockholders for the company included Prudential Insurance and Warner Communications.

In the congressional investigations in 1978, the use of street names drew substantial attention because of the close ties it allows between the financial community and power companies—in conflict with the spirit of the PUHC Act. While policies differ from one investment house to the next, in some cases the brokerage firms do vote the stock that elects a power company's directors, who oversee its policies and contracts. The specific policy that allows this is known as the New York Stock Exchange ten-day rule, under which the stockholder is given annual-meeting voting materials fifteen days before the deadline. If he or she does not vote within ten days, a brokerage firm may exercise the option to vote. In other situations the managers of large pension funds or trust funds vote on behalf of the multitude of participants.

Generally, the expansion of the power industry into other businesses, the growth of customer-ownership programs, and the concentration of control holds foreboding signs for consumers. The early warning signs are the threats of industry officials that the nation will see power shortages and blackouts if the dividend machines are not restarted and their plans for new construction approved.

Frank W. Griffith, a former chairman of the EEI and president of the Iowa Public Service Company, has said:

> This industry is built on investment, and unless investors perceive and receive an adequate return on their investment, there will be no further development of the industry. Utility managers across the country are saying publicly that they will no longer build facilities if the investor is not adequately compensated. . . . The vociferous few must not be allowed to direct the future of this industry or this nation. Regulators must summon the courage to face the future, however difficult and politically unpopular that task may be.

Similar to Energy Secretary Herrington's claim that there is a need to build as many as 200 new power plants or the nation will see power shortages, others have raised dire warnings. Robert Bigwood, chairman of the board of the Otter Tail Power Company in Minnesota, said the industry must be "unfettered from regulation in every reasonable way" in order to "clear the way for restoration and rapid development of nuclear energy. Until these things are done," he warns, "there is not going to be an energy-secure future for America."

Eugene Meyer claims he is uncertain of how all of this will come out. "The system is so shot," he says. "There is no doubt in my mind that in five years, or certainly in ten years from now, none of us will recognize the structure of the utility industry in terms of anything we knew historically." Asked if he thinks the industry might take the form of the massive generation and transmission companies foreseen by American Electric Power's Donald Cook and supported during the Nixon Administration, Meyer holds his cards close to his chest: "It won't be anything [like] what we have now. But I can't tell you what it will look like. I don't know." Griffith Morris, another vice-president for Kidder Peabody, told an industry gathering two months earlier, "We see a clear need to bring major nuclear power plants under construction on line. . . . Investment approaching a trillion dollars may be needed by the end of the century."

It's clear that what is being witnessed is not just a fight over the question of who will pay for the gargantuan losses hanging over the industry, nor is it just a fight over the rebirth of nuclear construction. What's in question is the future structure of the industry. Also at stake is a shift in the long-entwined institutional relationship of Wall Street and the electric industry. At bottom, the answer will determine far-reaching economic and political control that literally spreads like the electric systems into every community and every sector of the economy.

Some people believe that sympathetic federal officials like President Reagan, who had served as a spokesperson for General Electric for nearly a decade, Secretary of State George Shultz, who was on the board of J. P. Morgan's bank and president of the giant nuclear contractor Bechtel, former Secretary of the Treasury and White House chief of staff Donald Regan, who had been president of Merrill Lynch, and John S. R. Shad, a former E. F. Hutton executive and head of the SEC, will support restructuring of the industry and continuing evasion of public control. With new centralized holding company structures and weakened federal and state regulation, the power companies may be able to complete the troubled nuclear plants and pave the way for new development. But the wild card is what consumers will do when faced with the rising costs and new environmental threats that will come out of such events.

After more than a decade of fighting rising rates, intensified environmental threats, and the political chicanery of power companies, consumers will not take higher bills or power industry expansion lightly. What the nation witnessed in the massive consumer rebellion in the Northwest and the scattered beginnings of the "long and difficult struggle" over the $100 to $200 billion damage bill may be only the start of a larger clash on the horizon.

9. *The Coming Clash*

I believe we have reached a turning point in history. The antinuclear propaganda we are hearing puts democracy to a severe test. Unless the political trend toward energy development in this country changes rapidly, there may not be a United States of America in the twenty-first century.

Dr. Edward Teller,
father of the hydrogen bomb,
shortly after the Three Mile
Island accident in March 1979

The next decade of development of the electric industry will be central to the future of the United States. The question is whether monopoly control will continue through centralized power systems, or public challenges to the industry will encourage the growth of decentralized technology and allow the rise of diverse small power entrepreneurs and significantly increased energy efficiency. The answer is important not only to Wall Street and the concentration of economic power, but to regional economies, international resource issues, and environmental and social stability for Americans as well. The events unfolding in the mid-1980s promise a deepening turbulence.

The Soviet nuclear disaster at Chernobyl in April 1986 created worldwide shock—intensifying public fears of nuclear power. Within two weeks of the accident wind had carried radioactivity around the globe. In the northwestern corner of the United States radioactivity was picked up in rainwater. And in Europe large volumes of milk and vegetables were destroyed because of contamination. In an 18 mile radius around the Chernobyl plant, more than 86,000 people had been evacuated after a few days of uncertainty and confusion. Delayed and conflicting news reports stated that 35 people had received severe exposure and ten had died within the two weeks following the accident. Amid charges that the Soviets were maintaining a campaign of secrecy around the accident,

U.S. experts predicted rising cancer rates among some 100,000 people who lived in the vicinity of the plant.

Public opinion polls showed that a majority of Americans believed such an accident could also take place in the United States. An NBC/*Wall Street Journal* poll found 65 percent of the people interviewed were opposed to more nuclear construction in the United States. All of this indicated a hardening of the majority that had come to oppose nuclear power in the wake of the Three Mile Island accident in 1979.

Nevertheless, federal officials have remained firm in their commitment to support nuclear expansion. Curiously enough, at the same time U.S. spokesmen were criticizing the Soviets for their secrecy, the Nuclear Regulatory Commission was moving its deliberations behind closed doors by removing many of its meetings from requirements of the federal open meeting law. The Department of Energy, too, was tightening and centralizing its operations to better guide a nuclear resurgence.

The timetable for the Department of Energy's "second coming" of nuclear power included a first phase of clearing away "institutional barriers" and limiting the ability of citizens, by the late 1980s, to intervene. In 1986, major legislation advocated by industry leaders and Wall Street representatives was before Congress to amend the Atomic Energy Act of 1954 and streamline the licensing process, speed construction through advance site approval, and gain a rubber stamp for a new "safe" light-water reactor design. The two-pronged "second coming" plan aimed at having a new light-water reactor ready for commercialization by 1990, and a new breeder reactor design ready to go by 1995.

Internal Department of Energy documents showed a panel of industry representatives targeting the clearing of these barriers and the restoration of public confidence in nuclear programs through advertising and federal action. They recommended that President Reagan kick off the efforts with a major policy statement on nuclear power.

While consumer advocates and environmentalists fought the "second coming" plans in Washington, there was also turbulence in more than a dozen states where citizens and their state officials raised strong opposition to the siting of high-level radioactive waste dumps.

Creation of three such sites was mandated by the Nuclear Waste Policy Act of 1982, a law driven through Congress with strong backing from electric power companies and the Reagan Administration. The federal government's interest was in disposal of wastes from its atomic weapons program. The electric power industry wanted the sites so that the forty-year-old problem of what to do with nuclear waste from power plants can be declared "solved," clearing another major barrier to a second coming of nuclear construction in the early 1990s. The plan showed the depth of the disaster brewing.

One Department of Energy study for the dumps emerged four days after the re-election of President Reagan. It urged the creation of an "atomic priesthood" which would carry out a "ritual and legend" process to warn generations 10,000 years in the future of the danger of radioactive waste buried at the site. The waste would lie three thousand feet down, under a large triangular area bordered by raised mounds. At the center of the site, three twenty-foot-tall granite monoliths inscribed with warnings would stand on a concrete mat. It was labeled a "modern Stonehenge" site. Because our language may be incomprehensible three hundred generations from now, the fatal danger located underground would be communicated by stick-figure cartoons engraved on the monoliths. Other warnings might include a symbol resembling three sets of malevolent horns facing outward from a circle, or an undying artificial stench which people and animals would avoid. The study, produced by the Battelle Memorial Institute, said the "atomic priesthood" would reinforce these warnings with oral myths that threatened violators of the site with "some sort of supernatural retribution."

While the recommendations seemed to be the wildest cock-and-bull fantasy the federal government has ever paid tax dollars for, desperate solutions had to be found for the problem of what to do with radioactive waste which will present a danger for 100,000 years. Critics of the modern Stonehenge plan say that such a political solution is no substitute for a technological solution to radioactive wastes, which scientists are yet to agree upon. Like Eisenhower's waving of the nuclear "wand" in 1954 to break ground for the nation's first commercial reactor, the Stonehenge and atomic priesthood proposal emphasizes just how far federal officials are willing to go to promote the blueprint for a centralized nuclear future. It also illustrates the fact that political clout routinely substitutes for technological feasibility.

The Christian Science Monitor observed that the issue of trying to select the dump sites "may become one of the most potent federal-state issues since slavery." Although cities and towns in a number of states passed regulations outlawing the transport of radioactive materials through their boundaries, the local laws were pre-empted by the federal radioactive waste act. The waste act allows state governments and Indian tribes to have a consultant role, but gives final decision-making authority to Congress. The president is scheduled to select the nation's first permanent waste dump in 1991 and bring his recommendation to Congress for approval.

The prime candidates for the first waste dump site are the federal government's Hanford atomic weapons facility near Richland, Washington, the fertile farmland of Deaf Smith County, Texas, and Yucca Mountain, just outside the Nevada Weapons Testing site. A second permanent waste dump is being targeted for one of twelve eastern sites, including locations in Maine, New Hampshire, and Tennessee. As many as nine temporary dump sites where wastes would be kept for indefinite periods of time are also slated to be chosen. Ten states, including those in the waste transport corridors, have challenged the federal government in court. Federal officials have allegedly offered deals to keep the dumps out of certain states.

Governor Mark White of Texas has pledged that "sparks would fly" before a radioactive waste dump would be placed in Deaf Smith County. Nevada Governor Richard Bryan also promises opposition. And in Washington state, the nation's most pronuclear territory, the state legislature is split on the idea of a permanent dump at Hanford. Environmental and citizen groups are strongly opposing the idea. They point to a study on the basalt rock formations in the Hanford area and an assessment by the Nuclear Regulatory Commission that ground water could pick up and spread the radioactivity in as little as three hundred years, rather than the 81,000 years estimated by the Department of Energy. James B. Martin, an attorney with the Environmental Defense Fund, suggests that the Hanford area was chosen for its political support of nuclear ventures rather than for its geologic quality.

In Washington state and other target areas, the U.S. Committee for Energy Awareness is running sophisticated advertising and tracking polls to convince local populations of the safety of the sites. The industry group has also established and funded local business groups to neutralize local opposition. The Washington state group that supports siting of a permanent dump at Hanford is led by a former manager of the ill-fated WPPSS project, Robert Ferguson.

In Mississippi, James Palmer, Jr., an aide to Mississippi Governor William Allain, was extremely critical of the way the Department of Energy was withholding information about the selection process. He said, "Dealing with the DOE headquarters is about like our relationship with the Soviet Union—we meet, we greet, and we even eat, and then they cheat." Palmer described DOE as "the Kremlin on the Potomac," and said the department is telling states to "take the train, sit in the caboose, and keep your mouth shut." That's not what citizens are doing.

In Maine, as in other target states, citizens have staged a round of heated protests. In February 1986 a caravan of forty lumber trucks, well-drillers, and construction vehicles paraded in front of the Portland City Hall with their airhorns blaring. Inside, an overflow crowd holding a sea of signs which read "No!" listened to a Department of Energy official explain the reasons and the selection process for two sites in Maine. Governor Joseph Brennan and other state and local officials were highly critical. Brennan called the proposal for nuclear waste burial in the Lake Sebago area "unreasonable, unfair, and outrageous." Cliv Dore of the Passamaquoddy Indian Reservation charged that the Department of Energy "violated the law" when it failed to consult with the tribe over a potential dump at the reservation. Alva Morrison, a spokesperson for one of the citizen groups opposing the dump sites, told Department of Energy officials, "We can and will stop you. We have the power to do it and we are going to do it, not only in Maine, but everywhere you go in this country. If we can't stop you in Congress, we will stop you in the hills."

In addition to the conflict over siting the waste dumps, the rest of the nation will also witness tensions over the transport of highly radioactive waste on highways and rail lines. If all the waste from operating power plants is transported by truck, it will mean 6,405 shipments annually. If transport is done only by rail it will mean 830 shipments each year. Given the frequency of both railway and highway accidents, there is a certain degree of danger for communities all along the transport routes. The most heavily traveled roads that see radioactive cargo are Interstate 80, which spans the country, and Interstate 95, the north-south route from Maine to Florida.

The initial battles in Washington over the "second coming" conflicts regarding siting of the radioactive waste dumps and transport routes have added a new long-term dimension to the protests of the 1970s. In response, the power industry is connecting resolution of these issues with adequacy of power supplies. Power industry officials say that if new plants aren't built, there will be power shortages. And if dump selection is delayed, they will be forced to shut down plant operations.

J. Hunter Chiles III, an assistant to former Energy Secretary Don Hodel and now a business development manager for Bechtel Corporation, says that the nation is already in trouble for its future power supply. Because rate increases have been slowed and there is no long-term energy strategy for the nation, he says, utility planners are not making "purchasing decisions today for the power plants of the late 90s." Chiles says the problem is "our allowing people with single social issues to use the legal system to block anything: a new highway, a new building, a nuclear plant—you name it." Despite the fact that the nation had a power surplus of more than 30 percent in 1985, he predicts power shortages. "Eventually," he says, "[Americans will] see that we're short of electricity, and we'll go to expensive and short-range fixes—rationing, rotating blackouts, and high prices." The solution to avoid this crisis, in his view, is for decision makers to get off the backs of power companies and start pressuring regulators to allow higher rates and new construction. This vision to push back regulation is widespread within the industry and among Chile's colleagues who remain in the federal government.

In July 1985 Deputy Secretary of Energy Danny Boggs told the Senate Energy Committee, "The lights, in my personal view, will start to go out sometime in the 1990s and will do so increasingly as the decade progresses." He recommended that regulators allow companies to pass charges for canceled or partially constructed plants to consumers. Boggs was joined by major industry leaders, including William B. Ellis, chairman of the New England Power Pool and the chairman of Northeast Utilities, American Electric Power Company chief executive officer W. S. White, A. W. Dahlberg, president of Southern Company Services, Inc., and Floyd W. Lewis, chairman of Middle South Utilities. Lewis told the Senate Committee, "The greatest impediment to meeting the future demand for electricity is the regulatory climate," which he said penalized companies for taking the risks of building new plants, instead of shifting risk to the consumers. Each of these men said there would be future shortages if changes were not made and new building programs begun.

Alan Nogee, utility analyst for Environmental Action Foundation, called their tactics a new form of blackmail. "It's as if they have a protection racket going," Nogee said. "They tell families and business people that, if they pay for plants that aren't needed, then everything will be fine. Otherwise the lights will go out. Who are they kidding?"

Nogee agrees that electricity is vital to the nation's welfare, but points to the fact that the nation's use of electricity is vastly inefficient. "What the nation's electric systems need is growth based on 'least-cost' alternatives for power production. There are a range of options from energy-efficiency programs to renewable energy sources that the power companies are reluctant to implement because it cuts into their asset value and lowers the rates they can charge."

Similarly, Steven Ferrey, director of the energy project at the National Consumer Law Center, wrote to *The New York Times* attacking the prediction of power shortages: "We are literally awash in studies that prove conclusively that energy conservation in homes and factories, which frees existing power plants to serve new demand, is the cheapest way to an affordable energy future." Ferrey cited a Solar Energy Research Institute report (suppressed by the Reagan Administration) and a Library of Congress report, which conclude efficiency improvements could "generate" 32 percent to 100 percent of the growth in our electricity needs by the year 2000. "The [Reagan] Administration acts as if the growth of electricity demand is a natural force, beyond the power of man to control," Ferrey wrote. "In fact, growth of demand is a product of Government policies that encourage greater consumption and production, and discourage savings."

Essentially, the least-cost approach means looking at the business of selling electricity differently. It means encouraging efficiency and minimizing the need to build new plants—exactly the opposite of the traditional "build-and-grow" mentality that has dominated the industry. There are a variety of options in doing this. Utility companies can "buy" electricity from consumers, for example, by providing rebates for the purchase of energy-efficient appliances. This not only reduces residential bills and electricity demand but does it at a cost of only two cents per kilowatt hour. The saved electricity can then help to defer the need to build a new generating plant. Similar savings are available in commercial lighting and in electric motors, which make up two-thirds of industrial electricity demand.

For generating new electricity, cogeneration—using waste heat from industry to turn turbines—is the cheapest option at three to five cents per kilowatt hour. The federal Office of Technology Assessment estimates a potential for as much as 200,000 megawatts of cogeneration by the year 2000—an amount equal to 200 large nuclear plants. By comparison, new coal plants produce electricity at five to seven cents per kilowatt hour and new nuclear plants at ten to twenty-five cents per kilowatt hour.

A shift toward least-cost policies and efficiency and cogeneration would mean a further slowdown in growth of electric demand and some breakdown of monopoly control of the industry—a scenario that is not popular among power company executives.

Much of what will happen is being left to state and federal regulatory commissions. The signs are mixed at best. In California as much as 25 percent of the state's electricity could come from cogeneration, energy efficiency, and alternative energy sources by the year 2000. But the moratorium placed on new cogeneration projects and the elimination of renewable energy tax credits in 1985 may undermine this hope. Similarly, Public Service Company of New Hampshire officials have stated a need to halt the "gold rush" to small power development in New Hampshire to assure the market for centralized power plants such as Seabrook.

In other states the industry's political clout is proving substantial. Consumers in Ohio, for example, will pay a billion dollars in additional charges for the conversion of the abandoned Zimmer nuclear reactor to a coal-fired plant. "The three companies involved used the fear of blackouts to get what they wanted," economist Charles Komanoff said. Komanoff had worked out the original figures for an agreement on the conversion for the state of Ohio. A new coal plant would have cost $2 billion to $2.5 billion. Company executives at first agreed to keep costs down to this level and then came back with a figure of $3.6 billion, which allowed them to recoup $1.6 billion spent on the nuclear plant. "They told people that they would not exercise load management or due diligence if they did not get the full amount," Komanoff said. Company executives were also said to have made threats that they would discourage new industries from locating in their service areas, and blame the loss of jobs on state regulators and politicians.

In Illinois a similar situation has emerged, in which the Illinois Commerce Commission (ICC) has agreed to allow Sam Insull's old company, Commonwealth Edison, to provide discounted power rates to its 100 to 150 largest industrial customers. While this decision is cast as a move to stimulate business activity and jobs, company spokesman John Hogan admitted that the power company's sales have been sluggish. The new discounted rates will help to take up the existing slack, and with the new Byron and Braidwood nuclear plants coming on line, the five-year discounts will help to absorb the unneeded power. The catch is that it will also justify the need for the plants, perpetuate waste, and leave residential users with steady rate increases. Amory Lovins went before the ICC and testified that new energy efficient appliances, industrial motors, and bulbs could save as much power as the two Braidwood plants will provide, at a fraction of their cost. But his reasoning was turned aside. James O'Connor, the chairman of Commonwealth Edison, has warned repeatedly of a power shortfall in the 1990s without new nuclear and coal plants. In November 1985 he spoke at the joint annual conference of the Atomic Industrial Forum and the American Nuclear Society urging unity in regaining public confidence and working with the federal government.

At the federal level, officials such as Secretary John Herrington and his assistant Danny Boggs are working with the private power companies to see that they "control their own destinies." The keys to that destiny are fairly obvious. The Edison Electric Institute hopes to amend the Atomic Energy Act of 1954 to speed up the licensing and construction process for new plants and diminish opportunities for citizen intervention. The organization is also out to amend the Public Utility Regulatory Policies Act of 1978 to allow private power companies to gain ownership of small independent power facilities. They also intend to keep up efforts to alter the PUHC Act of 1935 to facilitate new holding company formation and place the power production under federal rather than state regulation. Taken together, these moves would allow the power empire to continue to dominate choices over the nation's energy future and dim the possibility of strongly increased efficiency and minimize competition from a horde of small entrepreneurs.

They also want to weaken public power and rural co-ops. In a significant weakening of public power systems, the Reagan Administration has given private power companies greater authority over power lines that cross federal lands—allowing them to refuse to transmit power destined for public or cooperative systems—an authority they have sought since the Eisenhower Administration scandals.

In opening the way for the private power companies to "control their own destinies," the administration has also raised the issue of selling off the federal power marketing agencies such as Bonneville Power Administration. Such a move would be a severe blow to many public power and rural cooperative systems that receive electricity from the federal agencies. In the Northwest the proposal has been met with staunch opposition and claims that it would rob the region of its lifeblood—cheap hydropower—and flatten the regional economy. Some critics fear that it would turn the Northwest into an "energy farm" for development in California. The proposal to sell the power marketing agencies is expected to surface repeatedly in the federal budget.

With such moves the Reagan Administration is not only ripping out the regulatory and economic balance placed in the industry during the New Deal but also squandering the cushion of the current power surplus and time that could be used to begin a national energy transition.

If state and federal regulatory agencies roll over to political pressure from the power industry and fail to push power companies into vigorous conservation and energy efficiency programs, the nation may lose the opportunity to make its electric systems more efficient, and to decrease the environmental risks and costs of electricity production. Power companies are already pulling older plants, which could be renovated, off line and dismantling them in order to decrease their surplus capacity and boost the need for new plants. Also, GE and Westinghouse are running programs to teach power system managers how to persuade their industrial customers to automate and electrify their plants. And industry ads and newspaper editorials have chimed in with the U.S. Committee for Energy Awareness to threaten blackouts and promote nuclear and coal expansion as the only viable option. In giving the private power companies control of their destinies, the Reagan Administration is also giving them control over a significant part of the nation's destiny.

Given the efforts to sweep aside public and regulatory control and the growing concern over the transport of radioactive wastes, acid rain, and steadily rising power costs, the turbulence of the last decade may be only a prelude to a deepening clash.

Although industry executives are aware of the possibility of heightened strife and the fact that it could "tear the economy apart," what they don't seem to acknowledge is that it will not be a confrontation between a "vociferous few" and power industry supporters. At the least, the crisis would bring paralyzing political debate and severe protests if the federal government supports industry interests over citizen concerns. Events of the last decade show that tens of thousands of consumers have been shocked into action to stop the runaway plans of the industry. And the slowly rising rates from the $100 billion to $200 billion damage bill are expanding that opposition further. The political clout of these consumers, adequately organized and backed by technological alternatives, could be more than a match for the power industry and its allies.

Already faced with the deepening environmental and economic risks and disillusioned by the failure of regulation, people in a number of communities are turning to an alternative to guarantee them local control—proposals for public takeovers of private electric companies. Their actions indicate that a drastically wrong turn was taken at the beginning of this century when private power companies were allowed to stifle the growth of local nonprofit public systems and dominate the nation's electric industry.

In New Orleans, San Francisco, Chicago, Long Island, and smaller communities scattered all across the nation, campaigns have surfaced to create new public power systems owned by consumers and run by local elected officials and professional managers. Operated as agencies of local government similar to water departments, the public system shifts the economic rationale from the industry's vision of maximum growth of a salable commodity to that of providing electricity as a nonprofit service and choosing what kind of growth is best for the community through public forums. Existing public power departments range from Seattle, Washington, to Kissimee, Florida. On average, the municipal systems provide electricity at rates 30 percent cheaper than private companies.

Private power officials usually claim this is due to the fact that public systems are tax-exempt. However, a 1984 study by the American Public Power Association (APPA) showed that public systems on average return a larger portion of their revenues to state and local government in lieu of taxes than do private companies. APPA also pointed out that while public systems pay no federal taxes, many private companies utilize tax loopholes to pay zero or minimal taxes to the federal government: In 1984, the private power companies were charged only $3.4 billion in taxes on total revenues of $154.8 billion—much less than half the average corporate tax rate of 20 percent.

Ideally suited to least-cost power alternatives, the public systems also take decisions over technology and power policies out of the hands of speculators and corporate investors and place control in the hands of consumers. But they also require diligent oversight by citizens. The reform of existing public power systems and creation of new public power systems may provide the only solid economic and political bridge for solar and conservation. It may also be the only practical way to avoid the deepening clash with private companies over the growing threats to the environment and economy. The battles now being waged to create new public power systems reveal how surprisingly little has changed in a century of development in the power industry and how the ruthlessness for control of electricity by private companies is still at the heart of this issue.

□ □ □

Perhaps nowhere has the issue become so starkly political as on Long Island. Questions over the safety and cost of the $4.5 billion Shoreham nuclear plant have thrown citizens and local officials against the Long Island Lighting Company (Lilco), its corporate allies, and federal agencies. The highly visible stream of events on Long Island are viewed by some individuals as a test for what may happen on a wider scale in the power industry.

For more than ten years the idea of taking over the Long Island Lighting Company has been brewing on Long Island. The focal point has been the company's high rates and Shoreham.

In September 1970 schoolchildren at the Joseph A. Edgar school in Rocky Point peered through the auditorium doors as the first hearings for the plant got under way. Local residents, nationally renowned scientists, and Nobel prize winners testified in opposition to the plant. It was hoped by environmentalists that Shoreham would be the test case on nuclear power. However, the case dragged on for two and half years in relative isolation. Finally in 1973 lingering questions and criticisms were turned aside and a construction permit was granted for work that had already been started. Nearly six years later, following the accident at Three Mile Island, the unanswered questions on the safety of the plant came echoing back and opposition to construction of the Shoreham plant broadened.

One of the most disturbing problems was that in the event of an accident at the plant, people living to the east halfway down the island would have to travel into the danger zone in order to evacuate. By late 1982, after failure to develop a workable evacuation plan, the issue came to a head. Suffolk County Executive Peter Cohalan called the plant a "billion-dollar mistake" and said it should never have been built. In a heated meeting of the county legislature, representatives vowed to never allow the plant to open.

Long Island Lighting Company officials were stunned. Costs for the plant were rising toward $3 billion in mid-1982 and the future of the company was tied to the fate of the plant. Company officials lobbied state and federal leaders to intervene, claiming that if Shoreham was not opened it would mean blackouts and bankruptcy. Nuclear Regulatory Commission Chairman Nunzio Palladino attempted expedited hearings for Shoreham's operating license, but other commissioners argued against this process. Angered by Palladino's action, Congressman Ed Markey twice asked Palladino to remove himself from the Shoreham issue, but Palladino refused.

In the meantime citizens had begun to see that Shoreham's rising costs would bring future rate increases. Along with safety concerns, they feared the impact of rate increases of 60 percent on Long Island's economy. In the summer of 1983, 28 industrial firms threatened to pack up and move off the island if Shoreham came on line with its predicted rate increases.

The Long Island Public Power Project, a group of local consumers and professional and union people, supported a feasibility study for a public takeover of the company that would leave Shoreham as a bad debt for the company's investors and lenders. They claimed that investors had taken a risk and the public should not have to cover their loss.

For two years officials remained deadlocked, while financing costs for the plant rose at the rate of $1.3 million per day. Finally, over the dissent of county and state officials, the Nuclear Regulatory Commission approved a low-power testing license for the plant on July 3, 1985.

That summer the takeover issue heated up, with both Long Island

Lighting Company and citizen opponents supporting candidates for the county legislature. In one revelation the Edison Electric Institute was found to have contributed $175,000 to a group called Citizens to Open Shoreham, which was running expensive advertising to reverse public opposition. But in late September, in the wake of the bruising winds of Hurricane Gloria, Long Island residents who were left in the dark without power for as long as a week turned increasingly bitter toward Long Island Lighting Company. On November 5, all eighteen legislators elected took stands in favor of a public takeover. Suffolk County Executive Peter Cohalan called the company "the big loser in the election."

In January 1986, Governor Mario Cuomo and both Republican and Democrat state senate leaders supported the takeover concept as the only way to break the deadlock over Shoreham and the plummeting fortunes of the Long Island Lighting Company. If the takeover proceeds, it could make a county power authority the third largest public power system in the United States, behind the Los Angeles municipal system and the Power Authority of Puerto Rico. Two different plans have emerged.

State officials favor a "friendly" takeover in which the state will finance the purchase of the company's stock. The power system would then be turned over to the Power Authority of New York, or a new Long Island Power Authority would be created to take over the system. This notion is supported by a recently formed citizen group, Citizens to Replace Lilco. But some consumer advocates are skeptical of this arrangement.

Marge Harrison, a member of a former governor's commission on public takeover and president of the Long Island Public Power Project, says that the "friendly" purchase would amount to a bailout for the company's investors and will jeopardize the potential for significant savings under public power. She points out that given its history of disregarding consumers, the state power authority could try and open Shoreham sometime in the future, and that new local municipal power systems with elected leaders need to be created to ensure consumer control. She favors standard municipal condemnation procedures that would allow the new power systems to take over only the assets that are "used and useful," thus significantly lowering the costs and providing a reduction in rates of about 30 percent.

Whichever course is followed, a lengthy court suit is expected to settle a price for the system, which is estimated now at $3 to $4 billion by the state and $18 billion by the company. The outcome could well affect public takeover efforts elsewhere.

□ □ □

In New Orleans, city councilors charged that New Orleans Public Service, Inc. (NOPSI), deliberately blacked out the city just days before a vote on whether the city would move to take over the company. In the early morning hours of Monday, January 21, 1985, with the temperature dipping to 14 degrees, controls reportedly froze at four of the company's generating plants. The power supply dropped dangerously low. In order to prevent a spreading region-wide blackout, company engineers began a controlled "rolling blackout," cutting power to sections of the city for an hour or two at a time as they struggled for more than eight hours to restore power. Fire broke out at a store where gas kept feeding into a heating system after electricity was cut. The temperature plummeted in homes that depended on electric starters or electric baseboards for their heating systems, and at least one person was alleged to have died as a result of the power loss.

That day, NOPSI officials appeared on local television, claiming they had averted a larger catastrophe and that if the controversial Grand Gulf nuclear plant had been on line, it would have saved the city from the rolling blackout. Critics and public officials charged that through carelessness and perhaps even purposeful negligence, the company had 50 percent of its generating plants shut down at the time the other four plants malfunctioned.

An investigation of the blackout by private engineers hired by the city revealed that NOPSI had promised to take steps to remedy cold-weather problems a year earlier in December 1983. However, in December 1984, the company mothballed and shut down for maintenance five of its plants—half of its generating capacity—at a time when there would be severe strain on the system, setting the stage for the blackout. The investigators also found that even if the Grand Gulf nuclear plant had been on line, transmission lines were overloaded, and its power would not have helped the city. Gary Groesch, spokesperson for Citizens for Independent Power, said, "I think this is the worst act that has ever been undertaken by a private power company." Brod Bagert, a local attorney and leader of the efforts to take over the power company, filed a $100 million class-action suit.

The catastrophe was only one of a series of events in the battle with NOPSI that had been going on for two years. It was being played out against the larger backdrop of Middle South Utilities holding company's long domination of the politics and economics of the region. The move to take over NOPSI—a subsidiary of Middle South—had begun in January 1983 after an announcement that rates would increase 73 percent to pay for contracts with the $3.4 billion Grand Gulf nuclear plant, a project owned by Middle South Energy, another subsidiary of Middle South Utilities. The rate hike meant a yearly increase of $114 million for the city through 1991. A few days after the revelation, an angry crowd showed up at a city council meeting to demand that the city buy out NOPSI under a franchise option it had with the company dating back to 1922. The city council appointed a twenty-two-person task force to explore the issue.

An initial study by the task force showed that the takeover proposal had merits. Under conditions originally set in the franchise, some anonymous legal genius had inserted a provision that the city could purchase the company for its rate base value (about one third of its market value) and that the takeover would not be subject to condemnation procedures, which in other cities often take years to resolve. The best news was that the city would not have to accept the company's contract for expensive power from the Grand Gulf nuclear plant.

The council voted $100,000 for a full feasibility study, which was done by the engineering firm of R. W. Beck. The study showed the city could save $250 million to $1 billion over a ten-year period by taking over the company and creating a New Orleans Public Power Authority. But there was more than just NOPSI to contend with. Middle South Utilities (MSU) holding company, which operates five power companies covering much of Arkansas, Louisiana, and Mississippi, saw the takeover effort as a bad precedent for other cities and towns in the region facing huge rate increases. It also posed a critical defeat for the plans of MSU's president, Floyd Lewis, to establish Middle South Energy as a regional generating company, under federal rather than state regulation.

On January 17, 1985, four days before the blackout occurred, the task force submitted a final recommendation to the city council, advising that takeover efforts begin. Floyd Lewis warned that a municipal takeover was "the worst thing that New Orleans could do." He promised to fight the effort. Lewis' opposition carries serious weight. Among its five subsidiary companies, MSU has 104 directors on its board, twenty of whom are also directors of major banks in the region—three from banks in New Orleans. But despite this opposition, the city council voted unanimously to create a New Orleans Public Power Authority.

MSU responded by organizing a Citizens Against Government Takeover committee, headed by George Denegre, an attorney for the power company. The company also enlisted its employees to hand out leaflets and put up campaign signs against the takeover on the lawns of neighbors and sympathetic businesses. The company also flooded the local radio stations and newspapers with advertising. Despite what was estimated to be an effort costing $1 million, on May 5, 1985, less than four months after the blackout, New Orleans residents voted 2–1 to take local regulatory control of the power companies, a first step toward a takeover. On the following day the confrontation heightened when New Orleans Public Service, Inc., and Louisiana Power and Light, the neighboring subsidiary of MSU, filed for the largest rate hikes in their history.

The rumor was that Middle South Utilities holding company would go bankrupt if it did not get the rate increases needed for Grand Gulf. More than ninety domestic and international banks had reportedly loaned MSU half of the $3.5 billion to construct the plant. As resistance to the rate increases mounted, Wall Street investment bankers were said to be crawling all over Arkansas, Mississippi, and Louisiana trying to convince political leaders that a bankruptcy for MSU would destroy the business climate and come back to haunt the region.

Ron Ridenhour, a reporter for the New Orleans weekly newspaper *Gambit,* documented attempts of a team of investment bankers led by Citicorp to halt the city's opposition to rate increases for Grand Gulf. Similar to threats that had emerged in the Northwest in the WPPSS confrontation, City Councilman Sidney Barthelemy told Ridenhour, "They told us the way we treated NOPSI was going to affect the credit for the entire city of New Orleans. . . . I guess you could have figured it as a veiled, sub-rosa threat. They insisted they weren't threatening us. They were just telling us the facts of life from their point of view."

What finally broke the crisis was the intervention of the Federal Energy Regulatory Commission. FERC allocated the rate increases for Grand Gulf between MSU subsidiaries in Louisiana, Arkansas, and Mississippi. Company stockholders were also made to take a loss of 12 percent, eventually rising to 22 percent. While the decision relieved the immediate pressure of bankruptcy, the sentiment for public takeovers spread from New Orleans to Arkansas.

A quick resolution is not expected. As events evolve, Long Island or New Orleans, or takeover efforts in Arkansas, could set vivid examples for other major cities. But the fight the nation will witness may be similar to one that has been going on in San Francisco for decades.

The efforts to create a public power system in San Francisco have been frustrated for more than seventy years in what Bruce Brugmann, editor of the San Francisco *Bay Guardian,* calls "one of the biggest and most costly ongoing scandals in U.S. history." Under the Raker Act of 1912, the city was given rights to dam the Hetch Hechty Valley in Yosemite National Park, some 130 miles away. The plan was to build the dam and take over that part of the Pacific Gas and Electric Company that served the city. But after the dam was built and a line strung ninety-nine miles toward the city, PG&E cut off the last thirty miles by building its own substation and power line extending under San Francisco Bay to the city. The city then faced the issue of whether or not to build a duplicate line under the Bay. As a result of a back-room city hall deal, the takeover was blocked. The city generated the power at Hetch Hetchy and sold it at wholesale rates to PG&E, which then sold it to the city's consumers at a higher retail price. Although the arrangement was declared illegal by the Department of the Interior in June 1923, enforcement measures were never carried out.

For the next two decades citizens attempted to pass votes for a takeover. A vote for a $2 million bond to purchase the company in 1925 went down to defeat by a vote of 59,727 to 52,216. From 1925 to 1941, PG&E defeated eight similar bond proposals. Despite the company's effective lobbying, the courts remained on the side of a takeover. In April 1941 the U.S. Supreme Court ruled that city officials had been illegally disposing of Hetch Hetchy power for fifteen years and that the Raker Act required a "publicly owned and operated power system" for San Francisco. Testimony in 1941 showed PG&E buying the power for $2 million and selling it back to consumers in the city for $9 million.

In the 1970s, rising electric rates brought the issue back to city politics. The power company responded by mounting strong opposition to block feasibility studies for a takeover. Brugmann says the company maintains its supporters in city hall and intimidates those who might oppose it. As in most major cities, the power company is a leading contributor in local elections and is interlocked with the city's most influential corporations.

At the front of PG&E's network is an influential group of major businessmen and bankers. "Of the 20 directors listed in PG&E's 1981 annual report," Brugmann says "six are directors or former directors of three powerful local banks with major PG&E holdings." M. Brock Weir, for example, is president of the Bank of California—PG&E's largest stockholder. Weir has not only opposed a public power system for San Francisco, but has also been involved in actions against existing public power systems.

In 1973 Weir was appointed head of the Cleveland (Ohio) Trust Company. There he was credited with threatening to withhold loans from the city unless it sold the long-embattled Cleveland municipal light system that Mayor Tom Johnson had fought to establish at the turn of the century. Mayor Dennis Kucinich refused. The showdown gained national recognition as the power industry poured contributions into an advertising campaign urging voters to sell the system. But Cleveland residents backed Kucinich and voted 2–1 to keep the system. Kucinich later said, "If I had cooperated with them and sold Muny Light to the private utility, everyone's electric rates would have automatically gone up. It would have set the stage for never-ending increases." In taking the issue to the ballot box, Kucinich said, "We gave the people a choice between a duly elected government and an unduly elected shadow government." Brugmann says Weir was chosen for the Cleveland job because of the success he had had in keeping the City of San Francisco from municipalizing the PG&E system.

Whether or not one believes in the image of a "shadow government" made up of powerful bankers and industrialists, the influence wielded by industry and the financial community is undeniably strong. In San Francisco, Brugmann says that in addition to Weir and other bankers who serve on the power company's board of directors under an exemption from the Federal Energy Regulatory Commission, PG&E executives serve as officers of various civic development committees and the city's Chamber of Commerce. The Byzantine corporate efforts to maintain private control over San Francisco's electricity have resulted in a number of scandals. In 1977, ties between the power company and city officials emerged when City Supervisor Robert Mendelsohn was found guilty by the California Fair Political Practices Commission of laundering campaign money. A key item in the case was a $12,000 no-interest loan to Mendelsohn from PG&E. Another incident surfaced in 1979 when City Supervisor Lee Dolson was scheduled to oversee a key hearing on a study to municipalize PG&E and was revealed as owning $9,800 of PG&E stock, an apparent conflict of interest that had gone unmentioned. City officials have also sided with the company to defeat proposals to take over PG&E.

The latest campaign to take over PG&E in 1982 was spurred by rising rates and $2 billion in cost overruns at the Diablo Canyon nuclear plant. Public power proponents experienced little trouble in collecting 20,000 signatures to place a proposal on the ballot calling for a feasibility study. An initial study by Accountants for the Public, a nonprofit organization of certified public accountants and academic specialists, had shown that the city could profit between $15 and $22 million annually by acquiring PG&E's distribution system. After bonds used to purchase the system are paid off, the figure would jump to $29 to $33 million annually, based on 1983 rates.

PG&E launched an extensive advertising campaign, claiming that a takeover would raise rates by 60 percent. Although the city government was required by law to provide an objective figure for the cost of the takeover, it declined to compile its own estimate and used the figure of $1.4 billion supplied by PG&E. Brugmann and others charged that the company used erroneous data in its calculations. After the election the company admitted its error, but in the meantime it stood by the figures which the city used. To counter the citizen drive for a feasibility study, the company also provided more than $220,000 to a front group, San Franciscans for Responsible Energy Policy. The company outspent proponents by five to one in the campaign, and the feasibility study was voted down. Although supporters were disappointed, Brugmann said it did a great deal to educate the people of San Francisco as to how PG&E is tied to city politics.

Elsewhere, the old notion of public power is also stirring renewed interest. While San Francisco's efforts have been frustrated, new public power systems were created in the California towns of Needles and Hayfork in 1983 and Trinity and Tuolumne counties in 1982. And in early 1986 the small lumber community of Susanville announced plans to take over the local distribution lines of CP National Corp. "CPN has outlived its usefulness," said Charles Richardson, a retired firefighter and an organizer of Citizen Power. The group hopes to create a new municipal system that will purchase electricity from local independent power producers such as lumber mills and geothermal plants. The town was pushed to find an alternative after CP National's rates increased by 30 percent in early 1985.

The shock of high costs from nuclear plants is also sparking takeover campaigns in a number of other states. In Page, Arizona, in January 1985, consumers stung by rising costs voted 1,574 to 149 to buy out the local distribution system from Arizona Public Service. Flagstaff, Williams, and Winslow are said to be watching the issue with great interest. In Utah, six towns were exploring takeovers of part of the Utah Power and Light company. In Kansas, several cities, still facing rate increases from the Wolf Creek nuclear plant, began exploring the idea. In Indiana, the city of LaPorte voted in November 1983 to create a public power system and take over part of Northern Indiana Public Service Company. Their effort was supported by a group of local industries seeking lower power costs and is being closely watched by the Governor's Committee on Indiana Utilities, which is examining the future of various utilities in the state in the wake of the $2.9 billion Marble Hill cancellations. Arkansas Governor

Bill Clinton has threatened to take over Arkansas Power and Light. Interest is also afoot in Connecticut, where citizens from a half dozen towns are examining takeovers of parts of United Illuminating and Connecticut Light and Power. And in New York in early 1985 a bill was introduced in the state legislature, with sixty-one cosponsors, that would facilitate a public takeover of all the private power companies in the state. Similar legislation in other states could be what breaks the takeover barriers set by private power in the early part of the century.

The surge of renewed interest in public power could have far-reaching effects in deflating the clash over private power's expansion plans and in providing a political bridge for conservation and solar technologies. In their 1981 book, *Power and Light: Strategies for the Solar Transition,* Richard E. Morgan and David Talbot speculated that the trend for such takeovers could become widespread. They hypothesized that private power companies facing bankruptcy and declining sales will be taken over first by nonprofit agencies. Then, as power demand remains low through the 1990s, because of independent conservation and solar and wind energy developments of home owners, local public systems would ultimately take over the nonprofit agencies. More likely now, soaring rates and growing threats of blackouts will be the force to drive an increasing number of communities to take over private power companies.

In a larger framework, the interest in public takeovers may return something of Roosevelt's yardstick function of public power to the industry. More importantly it could arrest the expansion of the economic and political influence of the private power empire and the future now being mortgaged to electrification of the economy under centralized power grids.

This growing trend for local nonprofit public power systems could have five general impacts. The first is a reduction of regulatory strife. This alone could save hundreds of thousands of man-hours and billions of dollars. Instead of being an arena for the perpetual conflict of corporate and public wills, regulatory agencies could function as planning bodies. Current rate and policy-making decisions could be made at the local level by elected officials, as they are in most of the present public power systems. Financing and participation in joint projects could be made subject to approval by voters, who through debate would be well aware of the risks and benefits of a particular project. State regulatory agencies studying and developing energy options could also work with regional planning bodies on scenarios for an entire region, as has recently been accomplished by a new Northwest Power Council.

A second impact of new public power systems would be taking control of electricity away from stock market speculators and financial firms and major corporations, which have a vested interest in boosting the sales of fuel, equipment, and financing. Financing the systems with revenue bonds, which have fixed rates of return, could provide greater stability to the stock market and opportunities for local people to invest their money in local ventures, which they'll be contributing to on the business side through their electric bills—keeping money in local and regional economies.

Third, as a nonprofit venture, public power development could follow least-cost policies and bring a jolt of life to small power producers, conservation businesses, and alternative energy firms, revitalize local economies, and provide an opportunity for the creation of the million jobs in energy services predicted by the Solar Energy Research Institute.

Fourth, placing local elected officials in charge of electric systems could help protect civil liberties and encourage greater participation in local government by providing direct citizen control over fundamental services. If an official elected to direct a local power system proves incapable or unresponsive, voters can remove him or her from office—a guarantee of efficient management that is now beyond the reach of customers of private power companies. Further, development of communities, automation of local industry, and electrification of heating systems or other extensive uses could be weighed by the community itself, rather than by corporate executives who may have priorities other than those of the community. On a larger scale, decisions made in such a public process can also take into account the hidden social and environmental costs of producing power.

Fifth, instead of being tied to a few large centralized generating plants, decentralized power systems linked into regional networks can provide greater stability, flexibility, and safety from sabotage or the effects of natural disasters. This would set the stage for production of power from photovoltaic cells or fuel cells, or other decentralized sources in the future. One scenario would be for a mix of small private power producers and public generating stations to feed into municipal distribution systems, such as the plan for Susanville, California. Consumer control and oversight over the system would help to assure implementation of least-cost policies and competition between power producers. At the regional level, nonprofit control and least-cost policies could increase cooperation in the power pools. This could help bring greater productivity and security to communities and more competitiveness to U.S. industry.

There are also international impacts in a slackened need for oil, gas, coal, and other imports, and for providing models of efficiency for nations whose electrical systems are in a stage of rapid development. In both developing countries and industrialized nations, a change in electric technology is already stirring.

In Sweden, where nearly all electricity is supplied by public power systems, a national vote was taken that declared a moratorium on construction of any new nuclear reactors and a phaseout of twelve operating reactors by the year 2010. In the interim the country has begun an extensive drive for energy efficiency and production of power from other sources. Austrians voted in 1978 not to activate their only reactor. In 1985 the Danish Parliament shelved plans to build nuclear plants. In France, a country which may be justifiably called the birthplace of electric lighting systems, the nationally owned power system is still committed to nuclear power, but is faced with on-going opposition and local referenda against the siting of radioactive waste dumps, particularly in the lake region 130 miles east of Paris. Insiders say that officials of the French nuclear system, which supplies 50 percent of the country's power and is often touted as the success story of the international nuclear industry, fear that flaws in the single design they used could bring about a call for shutdowns and cripple the country's electric output.

In Britain, Energy Secretary Peter Walker wants to move ahead with nuclear construction, but the industry is plagued by the resistance of people who don't want a reactor sited near their community. A Gallup poll taken in March 1986, shortly before the Soviet nuclear disaster, showed a majority of people opposed to further nuclear development. As in many other countries, British authorities also face opposition to the siting of radioactive waste dumps, an event that was forced after dumping of radioactive waste at sea was halted.

For developing nations of the Third World as well, a change may be afoot. Mexico, for example, pushed aside a plan to build twenty nuclear reactors in 1982, and China's plans to build two reactors have been followed by news that the Ministry of Water Resources and Electric Power believes the nuclear program will be unable to compete with coal-fired generation for several years.

In the Philippines, opposition has risen to the opening of the country's only nuclear reactor. Documents of former president Ferdinand Marcos show that Westinghouse made $11.2 million in payments to a Marcos associate in order to secure the contract for the plant. Westinghouse has acknowledged paying $17 million between 1976 and 1982 but denies any improprieties. A grand jury in Pittsburgh began investigating the payments in early 1986.

One of the only countries to ask for bids for new reactors is South Korea. The International Energy Agency, the World Bank, and other global institutions are attempting to turn the stall of reactor construction around, but the financial disaster in the United States looms as an unavoidable example of the trend toward technological dystopia. While international nuclear advocates argue that the "genie is out of the bottle," the truth is that no one can afford his services, and the promise of solar and other decentralized technology is more attractive to countries with no centralized transmission networks.

It may be that, in the future, we will only see nuclear plants being built in countries with highly centralized forms of government—nations which can force citizens to accept environmental risks and automatically pay the high costs of nuclear plants. If the power empire in the United States is successful with its threats of brownouts and blackouts, we may indeed witness growing centralization of our own government—already evidenced in the laws usurping local and state control to site waste dumps and efforts to have mandatory construction-work-in-progress charges, to place power production under federal rather than state regulation, and to amend the Atomic Energy Act of 1954 and limit citizen intervention.

In essence, a significant expansion of public power systems could put far-reaching decisions over a key part of the nation's future back in the hands of people at the local level. But public power should not be seen as a panacea. It's a system that carries its own set of problems. Statewide public power systems such as the New York Power Authority and the Washington Public Power Supply System can insulate appointed officials from public control and oversight. The same is true for the federal power marketing agencies such as Bonneville and the Tennessee Valley Authority. While these systems offer an advantage in being able to coordinate management and financing and development, without strong voter control they risk another form of centralization. But as is evident in the experience of consumers stopping construction of unneeded nuclear plants, this public centralization at its worst is more controllable than the closed-door operations of private companies. This fact will become especially vivid as regulatory authority erodes and private companies continue the trend toward holding company structures.

The best options are local nonprofit municipal systems and rural cooperatives run by elected officials with an open-door policy on information and voter control over investment in major programs and projects. What such consumer control does is establish the essential economic cornerstone on which everything rests—electricity as a nonprofit service. The political and business structure that rises from this foundation allows a process in which debate over development options can occur, with local people weighing social, economic, and environmental costs against the need for new power plants or conservation or other alternatives. Indeed, in view of the steady erosion of regulation, the transition to solar technology may need to be preceded by a twenty-five-year shift to the decentralized economic and political base local public power systems offer.

□ □ □

As the stakes rise and electricity becomes even more central to our lifestyle and industry, the fight for control of its generation, transmission, and distribution can be expected to grow increasingly bitter. The decline of oil production in the United States, anticipated to begin by 1995, will add another dimension to this struggle. The events of the past century show that the outcome of the struggle will have an effect far beyond simply providing electric power.

The actions and policies pursued by private utility executives over the past several decades to insure their control over the industry have left undeniable scars on the course of American history. Their push for state regulation legitimized the monopolization of electricity in a few hands and undermined the initial efforts at public control. The rise of their holding companies and uncontrolled financing and speculation helped fuel the stock market crash in 1929. Their struggle against federal support of local public power systems slowed rural electrification and left the nation short of electric capacity at the beginning of World War II. The rushed commercialization of nuclear power, which grew out of their drive to dominate this new energy source, has endangered communities with radioactive waste for centuries to come. Their undermining of regulation and promotion of unnecessary consumption and waste helped create the energy crisis of the 1970s and accelerated wide-ranging environmental problems. Their lobbying against air pollution laws has helped deepen the annual devastation caused by acid rain and the threat of the "greenhouse effect." And now the $100 to $200 billion damage bill for cost overruns and canceled plants and the threats to "dim America" if the bill is not paid is sending ripple effects through regional economies.

Throughout this time, costs have been padded and rates inflated in an unparalleled ripoff of American consumers. Energy efficient and solar technologies have been stifled. And out of a driven corporate self-interest that amounts to "friendly fascism," the nation's electric systems have been distorted by politics to the point that irrational atomic-priesthood proposals are seriously considered and feasibility studies to examine alternatives to decrease rates are fought against.

Taken together, the course of events over the past century, and in the last decade especially, provide warning signals of a need for dramatic change and evidence that the change will not come easily.

The scandals in San Francisco and New Orleans begin to show how entrenched the power empire has become in the business community and in local governments. The workings of regulation in New Hampshire and other states and the growing amount of power industry contributions to members of Congress show likewise the influence that private power companies and their network of supporters have over state and federal government.

While some alternative energy advocates believe that market forces and enlightened executives in the power industry will lead the way to a renewable energy transition, they fail to recognize this long history of institutional domination and how those institutions continue to shape the choices of even the most enlightened executives. In view of the statements of most officials, allowing the power companies to diversify into other lines of business and develop energy-service companies is likely to generate only a small amount of electricity and a great deal of grist for public relations programs.

It is not the failure of institutions that has brought this about as some would believe—but the fact that the institutions involved have functioned all too well. State regulation of the electric industry did the job Samuel Insull and others proposed it should do—it provided an alternative to public control and allowed the creation of huge private monopolies. Those monopolies pressed ahead with their purposes of maximum expansion and high profit. Wall Street provided abundant capital for the expansion and earned its bread-and-butter money from financing the private power companies. Equipment manufacturers and major construction contractors provided political support. And federal officials responded to the appeals of influential organizations and individuals to keep it all going.

As much as changing the local control and policies over electric power, what citizens are up against is changing this power empire that has wrapped itself in the nation's politics and regional economies. Such a challenge would be impossible if the industry had not pushed its influence into a glut of overbuilding and overfinancing. The $100 to $200 billion damage bill and the threats of brownouts and blackouts have made clearer than ever before the problems of how the industry operates and how much influence it truly carries.

Predicting the future is best left to those who read palms and not books, or history. But it's clear that a turning point has been reached in history. The events of the late 1980s and early 1990s will be critical to the environmental, political, and economic atmosphere of the next century. Whoever controls the policies over electric development in a given region will control an increasingly central part of its economy and politics.

In a more personal way, the turning point is one in which electricity is no longer the "miracle" it once was. For the generations that have grown up with it, the fire has lost its dazzle. The innocent belief in promises of technology producing power without cost is gone. The dream of abundance based on massive electric systems has given way to sobering economic and environmental realities. And the deepening social costs of trying to support infinite expansion have become all too visible. Fundamental change is inevitable. And the answer to what will come next is tied not only to technology, but the politics of a century-long "power struggle" that will shape the choices of technology.

In the past decade people have shown that they are going to demand greater corporate accountability and more responsiveness from public officials. In the changes to occur in the years ahead of us what happens in the local meeting halls will be as important as what happens in Congress or the board rooms of power companies. As citizens witness a threatened resurgence of nuclear power and the rise of new private power conglomerates, the words spoken by Pennsylvania Governor Gifford Pinchot in 1928 will take on new meaning:

> We need not be surprised that the State and Federal authorities have stood in awe before this gigantic nationwide power monopoly, because beside it, as its creator, financial supporter, and master, stands the concentrated money power of the United States, which today is the dominating money power of the world.
>
> This dominating money power has seized upon the electric power resources of our country with the full realization of the fact that before very many years it will be not the "hand that rocks the cradle" but the hand that turns the electric switch that will rule the land.
>
> Therefore the electric power monopoly deserves the fullest public attention. The people ought to know what it is and why it is and how it affects them. All the facts about it ought to be publicly available either through government agencies or private effort. The people must learn to judge intelligently of its advantages and its evils. Everything about it should be investigated fearlessly and published fully, because we must learn to regulate and control it before it smothers and enslaves us.

Notes

Chapter 1 Warning Signs

PAGE

1 Much of the material on events surrounding the collapse of the Washington Public Power Supply System contained in this chapter was gathered in interviews, documents, and personal reporting under a grant from the Fund for Investigative Journalism. Dorothy Lindsey's statement is found in P. J. Rader, "Electric Rates Spark Angry Turnout in Hoquiam," *Seattle Times,* February 15, 1982.

2 Information on the "second coming" is from the *Strategic Plan for the Civilian Reactor Development Program* (draft), U.S. Department of Energy, October 25, 1985, and additional documents collected under the Freedom of Information Act.

Energy Secretary John Herrington's support of a second coming can be found in his public speeches, such as one before the Nuclear Power Assembly in Washington, D.C., on May 8, 1985 (Washington, D.C.: Department of Energy press office). Nuclear Regulatory Commissioner Lando Zech's positions are also expressed in his public statements; see "Remarks by Commissioner Lando Zech, Jr., at Edison Electric Institute Governmental Affairs Conference, Marco Beach Hilton," Marco Beach, Florida, January 16, 1986. See also Warren Richey, "Reagan Plans Top-Level Review of Nuclear Power, Seeking to Revive Industry," *Christian Science Monitor,* May 10, 1985.

Power industry executives testified before the Senate Energy and Natural Resources Committee on July 23 and 25, 1985. See also "Utility Executives Reveal Shortages, Senate Listens," *Electrical World,* September 1985, pp. 17–22.

3–9 For more background on the evolution of the WPPSS default see Scott Ridley, "Money Meltdown: How a Nuclear System Called Whoops Went Critical," *The New Republic,* August 29, 1983. See also "Special Report," *Seattle Post-Intelligencer,* February 13, 1983, and ongoing coverage provided by Northwestern dailies from November 1981 through July 1983.

PAGE

6 The nine-page memo on the meeting of 21 business and industry representatives on July 23, 1982, was written on the stationery of Preston, Thorgrimson, Ellis and Holman, a prestigious Seattle law firm that represented Direct Service Industries, a trade association for large industrial electricity consumers. Eric Redman, a former partner of the firm, was noted as the author. Redman denies knowledge of the memo.

7 Charles Komanoff's comments were made in an interview on April 26, 1984. See also similar statements made by Komanoff in "Are Utilities Obsolete?" *Business Week,* May 21, 1984, pp. 116–129.

8 Alan Nogee's analysis of the magnitude of economic damage from the cost of nuclear plant construction is contained in the report *Rateshock* (Washington, D.C.: Environmental Action Foundation, October 1984).

See *Forbes*'s cover story on the crisis over nuclear power, "Nuclear Follies," by James Cook, February 11, 1985.

9 The figure for water use by thermoelectric nuclear plants is from "Estimated Use of Water in the United States in 1980," U.S. Geological Survey (Washington, D.C.: U.S. Government Printing Office). The estimate includes both fresh and salt water.

9–11 See the source material in Chapters 2 through 5, from which this material is condensed and summarized.

12 Amory Lovins's statement alluding to "friendly fascism" can be found in "A Neo-Capitalist Manifesto: Free Enterprise Can Finance Our Energy Future," *Politicks & Other Human Interests,* April 11, 1978.

14 Pennsylvania Commissioner Michael Johnson's statement was made in an interview on August 15, 1984.

Nuclear Regulatory Commissioner Victor Gilinsky's comment can be found in "Atomic Energy Industry Jolted Anew," *Washington Post,* January 28, 1984.

The correlation between congressional voting records and political action committee contributions from the nuclear industry can be found in the report "Nuclear PAC's—$3 Million Votes," *Public Citizen,* July 1984.

15 Tax policy analysis can be found in a report by H. Richard Heede, Richard Morgan, and Scott Ridley titled *The Hidden Costs of Energy* (Washington, D.C.: Center for Renewable Resources, October 1985).

16 The statement on a potentially serious conflict between consumers and the electric industry is contained in Joseph Dukert, *A Short Energy History of the United States* (Washington, D.C.: Edison Electric Institute, 1980).

See *Strategic Plan for the Civilian Reactor Development Program*

PAGE

and speeches previously cited as well as "Address of John S. Herrington, Secretary of Energy, to the Atomic Industrial Forum/American Nuclear Society," San Francisco, California, November 13, 1985.

17–18 The statements by Amory Lovins are from an interview, July 8, 1984. See also Lovins's testimony before the House Subcommittee on Energy Conservation and Power, June 26, 1984, and his analysis of energy efficiency presented in his books, such as *Soft Energy Paths* (Cambridge, Mass.: Ballinger Publishing Company, 1977).

18 North American Electric Reliability Council estimates for cogeneration and power needs can be found in *1985 Reliability Review* (Princeton, N.J.: North American Electric Reliability Council, 1985).

19 Tina Hobson's statement was made in a press luncheon speech on Capitol Hill on October 16, 1985.

20 Suffolk County Legislator Wayne Prospect made the comments on preventing operation of the Shoreham plant during a meeting of the Suffolk County Legislature on February 17, 1983. Chicago Mayor Harold Washington's statements can be found in "Chicago May Pull the Plug on Edison," *Chicago Tribune,* October 11, 1985.

Chapter 2 Behind the Miracle and the Myth

PAGE

22 The description of the dramatic scene in the Cleveland City Council Chamber is found in the *Cleveland Plain Dealer,* February 6, 20, 1905, and Eugene C. Murdock, "Life of Tom Johnson" (doctoral dissertation, Columbia University, 1951), pp. 256–257.

23 Mayor Tom Johnson's statements are from *Tom Johnson, My Story* (New York: D. W. Huebsch, 1951), pp. 26–37, 194.

24 The first public street lighting demonstration in America is described in Mel Gorman, "Charles F. Brush and the First Public Electric Street Lighting System in America," *Ohio History,* vol. 70, 1961, pp. 28–144, and William Ganson Rose, *Cleveland: The Making of a City* (New York: World Publishing Company, 1950), p. 423.

The quotation describing Wabash residents' reaction to street lighting is found in Vic Reinemer, "Public Power's Roots," *Public Power,* centennial issue, September–October 1982, pp. 22–23.

32–33 The quotations by Hazen Pingree are found in Charles N. Glaab and Theodore A. Brown, *A History of Urban America,* 2nd ed. (New York: Macmillan, 1976), p. 198, and Melvin G. Holli, *Reform in Detroit* (New York: Oxford University Press, 1969), pp. 80–81, 84.

34 Patrick Collins's statement concerning the general lighting bill in Massachusetts is taken from "Municipal Gas and Electric Plants in Massachusetts," *Political Science Quarterly,* vol. XVII, p. 33.

PAGE

37–38 Sources for the early history of the Cleveland municipal lighting plant include Edward J. Kenealy, *The Cleveland Municipal Lighting Plant* (Cleveland, Ohio, 1935), pp. 48–50; Cleveland Municipal Lighting Department, *Bulletin I,* April 1915, pp. 18–19; and Carl D. Thompson, "Municipal Electric Light and Power Plants in the United States and Canada," *Public Ownership League Bulletin No. 1,* 1922, pp. 79–80.

38–39 Samuel Insull's statement regarding natural monopolies and the need for regulation is found in the *National Electric Light Association, Proceedings,* 1898, p. 27.

40 The differing views of state regulation are found in David Nord, "The Experts Versus the Experts: Conflicting Philosophies of Municipal Utility Regulation in the Progressive Era," *Wisconsin Magazine of History,* vol. 58, 1975, pp. 219–236.

40–41 The quotation attributed to a consulting economist, Ernest Bradford, is found in William E. Mosher, ed., *Electrical Utilities: The Crisis in Public Control* (New York: Harper & Brothers, 1929), p. 24.

41–42 Samuel Insull's story and the development of the holding company structure are described in Forrest McDonald, *Insull* (University of Chicago Press, 1962); Frederick Lewis Allen, *Lords of Creation* (New York: Harper & Brothers, 1935), pp. 266–303; M. L. Ramsay, *Pyramids of Power: The Story of Roosevelt, Insull and the Utility Wars* (New York: Bobbs-Merrill Company, 1937); and Bernard Ostrolenk, *Electricity for Use or for Profit* (New York: Harper & Brothers, 1936).

44 Statistics concerning the number of publicly owned systems purchasing their total electrical output from private companies are found in Herbert B. Dorau, "Rates in the Electric Industry under Municipal Ownership," *Annals,* vol. 201, p. 41, and Paul J. Raven and Marion Sumner, *Municipally Owned Utilities in Nebraska* (Chicago: Institute for Economic Research, 1932), pp. 57–58.

The figures for water power generation and the extent of private control are taken from *Historical Statistics of the United States: Colonial Times to 1970,* part 2 (Washington, D.C.: U.S. Department of Commerce, 1975).

45–46 Sources for the history of the National Popular Government League and Public Ownership League include Bethany Weidner, "Public Power's Early Evangelists," *Public Power* (November–December 1983), pp. 70–74; *Brief History of the Public Ownership League and What It Has Done to Protect and Promote Municipal and Public Utilities and Natural Resources* (Chicago: Public Ownership League, 1929); and Judson King, *The Conservation Fight: From Theodore Roosevelt to the Tennessee Valley Authority* (Washington, D.C.: Public Affairs Press, 1959).

PAGE

48–51 Quotations concerning the private power industry's propaganda campaign are found in Ernest Gruening, *The Public Pays . . . and Still Pays: A Study of Power Propaganda* (New York: Vanguard Press, 1931), pp. 98–99; Mosher, op. cit., pp. 158–160; and William Rodgers, *Brown Out: The Power Crisis in America* (New York: Stein & Day, 1972), p. 81; Ramsay, op. cit., p. 128.

51 The excerpt from J. F. Owens's speech "Electric Night" is found in Ernest Gruening, "Power as a Campaign Issue," *Current History,* vol. 37, October 1932, pp. 223–224.

53 The contemporary Wall Street analyst description of holding company deals is found in William Z. Ripley, *Main Street and Wall Street* (Boston: Little, Brown & Company, 1927), p. 303.

54 Statistics concerning holding companies' control of the nation's electricity are found in FTC, *Utility Corporations, 70th Congress 1st Session,* Senate Document no. 92, part 72-A, pp. 37–40.

Figures for utility regulation expenses are taken from Mosher, op. cit., pp. 13–14, 25–26, and H. S. Raushenbush and Harry W. Laidler, *Power Control* (New York: New Republic, 1928), pp. 118–119.

55 J. F. Owens's statement concerning the birth of the electric industry is found in J. F. Owens, "President's Address," *NELA Bulletin,* October 1932, p. 8. Albert Einstein's remark on the Golden Jubilee of Light is found in Thomas W. Martin, *The Story of Electricity in Alabama* (privately printed, 1949).

56 The description of the breakup of the National Electric Light Association and the establishment of the Edison Electric Institute is found in "Convention Called to Dissolve NELA: Edison Electric Institute to Replace the Association," *NELA Bulletin,* February 1933, p. 59.

Besides the above material, the following general accounts were particularly useful for documenting the early history of the power industry: Everett W. Burdett, *The Agitation for Municipal Ownership in the United States—Its Origins, Meaning and Property Treatment* (New York: National Electric Light Association, 1906); Herbert B. Dorau, *The Changing Character and Extent of Municipal Ownership in the Electric Light and Power Industry* (Chicago: Institute for Research in Land Economics and Public Utilities, 1929); Thomas P. Hughes, *Thomas Edison: Professional Inventor* (London: Her Majesty's Printing Office, 1976); David Morris, "The Pendulum Swings Again: A Century of Urban Electric Systems," in *Decentralizing Electricity Production,* Howard Brown, ed. (New Haven: Yale University Press, 1983); Frank Parsons, *The City for the People* (Philadelphia: C. F. Taylor, 1901); Gifford Pinchot, *The Power Monopoly: Its Makeup and Its Menace* (Milford, Pa., 1928); *Central-Station Electric Service: Its Commercial Development and Economic Significance as Set Forth in the Public Addresses (1897–1914) of Samuel Insull* (Chicago: pri-

PAGE

vately printed, 1915); Samuel Insull, *Public Utilities in Modern Life: Selected Speeches, 1914–1923,* William E. Keily, ed. (Chicago: privately printed, 1924); Thomas P. Hughes, *Networks of Power* (Baltimore: Johns Hopkins University Press, 1983); Sheldon Novick, "The Electric Power Industry," *Environment* (November 1975); Ronald Clark, *Edison: The Man Who Made the Future* (New York: G. P. Putnam & Sons, 1977); and Harold C. Passer, *The Electrical Manufacturers, 1870–1900: A Study in Competition, Entrepreneurship, Technological Change and Economic Growth* (Cambridge: Harvard University Press, 1953).

For the early history of regulation see James C. Bonbright, *Public Utilities and the National Power Policies* (New York, 1940); Paul J. Garfield and Wallace Lovejoy, *Public Utility Economics* (Englewood Cliffs, N.J.: Prentice-Hall, 1964); Richard Hellman, *Government Competition in the Electric Utility Industry: A Theoretical and Empirical Study* (New York: Praeger, 1972); Forrest McDonald, "Samuel Insull and the Movement for State Utility Regulatory Commissions," *Business History Review,* vol. XXXII, Autumn 1958, pp. 241–254; *Municipal and Private Operation of Public Utilities: Report of the National Civic Federation,* 3 vols. (New York: National Civic Federation, 1907); and "Proceedings of the Conference of American Mayors on Public Policies as to Municipal Utilities," *Annals,* vol. LVII (American Academy of Political and Social Science, 1915).

Among the many works focusing on the early history of public power at the state and local level the following were especially helpful: Frederick Bird and Frances Ryan, *Public Ownership on Trial* (New York: New Republic, 1930); Robert E. Firth, *Public Power in Nebraska: A Report on State Ownership* (Lincoln: University of Nebraska Press, 1962); Edmond Lincoln, *The Results of Municipal Electric Lighting in Massachusetts* (Boston: Houghton Mifflin, 1918); and Ernest Bradford Smith, *Municipal Electric Lighting* (Madison: Wisconsin Free Library Commission Legislative Reference Department, 1912).

Chapter 3 The Turbulent Thirties

PAGES

57–59 Source material for the collapse of Samuel Insull's empire includes McDonald, op. cit., pp. 274–304; N. R. Danielian, "From Insull to Injury: A Study in Financial Jugglery," *Atlantic Monthly,* vol. 141, April 1933, pp. 497–508; Donald Richberg, "Gold Plated Anarchy: An Interpretation of the Fall of the Giant," *The Nation,* vol. 136, April 5, 1933, pp. 368–369; and Norman Thomas, "Owen D. Young and Samuel Insull," *The Nation,* vol. 136, January 11, 1933, pp. 35–37.

59–60 James C. Bonbright's statement concerning how holding companies have changed over the years is found in *New York Times,* April 10, 1932.

PAGE

60 The quotation attributed to an executive concerning the public reaction to the collapse of Insull's holding companies is found in Marquis W. Childs, "Samuel Insull: The Collapse," *New Republic,* vol. 72, October 5, 1932, p. 203.

61–62 Figures for power industry income during the Depression are found in Ramsay, op. cit., pp. 78–83, and Nicholas B. Wainwright, *History of the Philadelphia Electric Company, 1881–1961* (Philadelphia: Philadelphia Electric Company, 1961), p. 238.

62–63 The statements signed by 37 congressional leaders concerning the "power question" are found in Ernest Gruening, "Power as a Campaign Issue," *Current History,* vol. 37, October 1932, p. 44.

63–64 Background material on Newton D. Baker's candidacy in 1932 is found in Oswald Garrison Villard, "Newton D. Baker—Just Another Politician," *The Nation,* vol. 134, April 13, 1932, pp. 414–418, and C. H. Cramer, *Newton D. Baker: A Biography* (New York: World Publishing Company, 1961), pp. 242–243.

64 Judson King's statement concerning the "power fight" at the Democratic convention is found in Judson King, "Roosevelt's Power Record," *New Republic,* vol. 72, September 7, 1932, pp. 96–98.

66 President Herbert Hoover's remarks concerning federal operation of the power plant at Muscle Shoals are found in Arthur A. Schlesinger, Jr., *The Crisis of the Old Order, 1919–1933* (Cambridge: Riverside Press, 1957), p. 121.

67–68 Franklin D. Roosevelt's "Portland speech" is reprinted in Samuel I. Rosenman, comp., *Public Papers and Addresses of Franklin D. Roosevelt,* vol. I (13 vols., New York, 1938–50), pp. 734–740.

69 George Norris's recollections of the "irreconcilable conflict" over the Muscle Shoals project are found in George Norris, *Fighting Liberal: The Autobiography of George Norris* (New York: Macmillan, 1945), pp. 245–246.

70–72 Quotes and a historical narrative of the private power industry's opposition to the TVA are found in Arthur M. Schlesinger, Jr., *The Coming of the New Deal* (Cambridge: Riverside Press, 1965), pp. 325–326, and Tom K. McCraw, *TVA and the Power Fight, 1933–1939* (Philadelphia: J. B. Lippincott Company, 1971), pp. 35–36.

72–73 The description of the meeting between David Lilienthal and Wendell Willkie is found in *Journals of David E. Lilienthal,* vol. I: *TVA Years, 1932–1945* (2 vols., New York: Harper & Row, 1964), pp. 711–712.

PAGE

73 A description of the Alcorn County Electric Power Association is found in "The First Co-op in the Valley: An Experiment Triumphs," *Rural Electrification,* July 1983, pp. 19–20.

73–74 The Chattanooga fight is described in Hellman, op. cit., pp. 176–177, and McCraw, op. cit., pp. 128–130.

76–78 Quotes and historical narrative for the private power industry's lobbying campaign against the Public Utility Holding Company Act are found in Philip J. Funigiello, *Toward a National Power Policy: The New Deal and the Electric Utility Industry, 1933–1941* (University of Pittsburgh Press, 1977), pp. 98–120, and Arthur A. Schlesinger, Jr., *The Politics of Upheaval, 1935–1936* (Cambridge: Riverside Press, 1966), pp. 302–324.

79–80 Harold Ickes's statement concerning the role of private utilities in promoting rural electrification is found in D. Clayton Brown, *Electricity for Rural America: The Fight for the REA* (Westport, Conn.: Greenwood Press, 1980), p. 40.

80 George Norris's boyhood memory of chores done in the "flickering light of a coal-oil lamp" is found in *The Fighting Liberal,* pp. 318–319.

Sources for the early history of rural electrification include Brown, chaps. 1–7, *Electric Power and Government Policy: A Survey of the Relations between the Government and the Electric Power Industry* (New York: 20th Century Fund, 1948); Donald H. Cooper, ed., *Rural Electric Facts, American Success Story* (Washington, D.C.: National Rural Electric Cooperative Association, 1970); and Clyde T. Ellis, *A Giant Step* (New York: Random House, 1966).

81–82 Harry Slattery's statement concerning power industry tactics and their effect on rural electrification in the 1930s is found in Harry Slattery, *Rural America Lights Up* (Washington, D.C.: National Home Library Foundation, 1940), pp. 115–121.

82–83 Quotes describing the first rural cooperative members' enthusiasm are found in "The First Co-op in the Valley," op. cit., pp. 19–20.

84 The description of the Public Utility District Movement is taken from Gus Norwood, "Public Power's Roots Deep in the Northwest," *Public Power,* January–February 1982, pp. 108–114.

84–86 For a description of the struggle over wartime preparedness see Funigiello, op. cit., pp. 226–271; Eliot Janeway, *The Struggle for Survival* (New Haven: Yale University Press, 1951), pp. 125–126, 234–235; and Harold L. Ickes, *The Secret Diary of Harold L. Ickes: The First Thousand Days,* vol. III. (3 vols., New York, 1953–54), pp. 578–579, 586–88, 604–5, 607.

In addition to the above citations the following general accounts were useful in preparing the sections on the Muscle Shoals contro-

PAGE

versy and the early history of the TVA: Preston J. Hubbard, *Origins of the TVA: The Muscle Shoals Controversy, 1920–1932* (Nashville: Vanderbilt University Press, 1961); North Callahan, *TVA Bridge over Troubled Waters* (South Brunswick and New York: A. S. Barnes, 1980), chaps. 1–4; Erwin C. Hargrove and Paul K. Conkin, *TVA: Fifty Years of Grassroots Bureaucracy* (Urbana: University of Illinois Press, 1983); David E. Lilienthal, *TVA: Democracy on the March* (New York: Harper & Row, 1953); Roscoe C. Martin, *TVA: The First Twenty Years: A Study in the Sociology of Formal Organization* (Knoxville: University of Tennessee Press, 1956); C. H. Pritchett, *Tennessee Valley Authority: A Study in Public Administration* (Chapel Hill: University of North Carolina Press, 1943); and Philip Selznick, *TVA and the Grassroots* (Berkeley: University of California Press, 1949).

Among the dozens of studies of Franklin D. Roosevelt the most useful were Frank Freidel's *The Triumph* and *Launching the New Deal* (Boston: Little, Brown, 1956 and 1973). Roosevelt's gubernatorial experience with the power issue is explored in Bernard Bellush, *Franklin D. Roosevelt as Governor of New York* (New York: Columbia University Press, 1955), pp. 208–268. For Wendell Willkie's thoughts on government ownership of the power industry see *Free Enterprise: The Philosophy of Wendell L. Willkie as Found in His Speeches, Messages and Other Papers* (Washington, D.C.: National Home Library Foundation, 1940); *The Public Utility Problem: Its Recent History and Possible Solution.* Address before the Bond Club of New York, December 19, 1935, New York City; and Justin R. Whiting, *Wendell L. Willkie: Courageous Pioneer of the Utility Industry* (New York: Newcomer Society in North America, 1950).

For the early history of rural electrification see also Funigiello, op. cit., pp. 122–173; Marquis W. Childs, *The Farmer Takes a Hand* (New York: Doubleday, 1952); M. L. Cooke, "The Early Days of the Rural Electrification Idea: 1914–1936," *American Political Science Review,* vol. 42, June 1948, pp. 431–447; and H. S. Person, "The Rural Electrification Administration in Perspective," *Agricultural History,* April 1950.

Chapter 4 The Coming of Nuclear Power

PAGE

87 The statement by Paul F. Genachte of the Chase Manhattan Bank is from a speech he delivered to the Atomic Industrial Forum annual conference held in Chicago, September 25–27, 1956; see *Atomic Industrial Forum 1956 Conference Proceedings.*

88 For H. G. Wells's visions of atomic energy see H. G. Wells, *The World Set Free* (New York: E. P. Dutton, 1914), p. 51.

89 Robert Bacher's statement and the statements of many other scientists optimistic about the uses of atomic energy in the late 1940s and early

PAGE

1950s can be found in contemporary issues of the *Bulletin of Atomic Scientists.* This statement is contained in the March 1949 issue, p. 80.

There are a number of books on the period of atomic development after World War II. Among those that provide a glimpse of the inner workings of government and industry are Corbin Allardice and Edward Trapnell, *The Atomic Energy Commission* (Praeger Publishers, 1974); Harold Orlons, *Contracting for Atoms* (Brookings Institute, 1967); Lewis L. Strauss, *Men and Decisions* (Doubleday, 1962), and the reports of the Joint Committee on Atomic Energy and the Atomic Energy Commission's annual reports to Congress.

90 David Lilienthal's statement on the broad dilemma facing mankind over the development of new technologies appeared in "Science and Man's Fate," *The Nation,* July 13, 1946.

93 For the full transcript of the confirmation hearings for the first members of the Atomic Energy Commission, see "Confirmation Hearings for the Atomic Energy Commission and General Manager," Senate section of the Joint Committee on Atomic Energy, January, February, and March 1947.

94 For information on the nationalization of electric utilities in France see *Electric Power in France,* by the Chief Engineer of the European Theater, 1946 (Library of Congress). For Great Britain's nationalization see Sir Henry Self and Elizabeth M. Watson, *Electric Supply in Great Britain* (London: Allen and Unwin, 1952).

95 American Gas and Electric Company head Philip Sporn's comments on the unprecedented power industry expansion can be found in *Bulletin of Atomic Scientists,* October 1950, p. 306.

95–96 Source material for the Republican effort to change the direction of federal power policy in the post–World War II era and the Fulton steam plant controversy is found in Susan M. Hartman, *Truman and the 80th Congress* (Columbia: University of Missouri Press, 1971); John R. Waltrip, *Public Power During the Truman Administration* (doctoral dissertation: University of Missouri, 1965); Aaron Wildavsky, *Dixon-Yates: A Study in Power Politics* (New Haven: Yale University Press, 1962); and Bonnie Baack Pendergass, *Public Power, Politics and Technology in the Eisenhower and Kennedy Years* (New York: Arno Press, 1979).

96 Truman's remarks during the 1948 presidential campaign are found in *The Truman Program: Addresses and Messages of President Harry S. Truman* (Washington, D.C.: Public Affairs Press, 1948, 1949), pp. 227–228.

97 The uproar over copper hoarding and allegations that the War Production Board was favoring private power companies are contained

PAGE

in D. Clayton Brown, *Electricity for Rural America* (Westport, Conn.: Greenwood Press, 1980), pp. 84–85.

98 The quotation concerning the private power company interest in stressing socialist themes in the fight against public power is found in Virginia Reid, "Needed: New Power Company Propaganda Probe," *Public Power,* May 1960, pp. 7–8.

99 The skepticism about rapid development of nuclear power expressed by Enrico Fermi and Robert Oppenheimer can be found in *Bulletin of Atomic Scientists,* October 1950. The equally sober appraisal by the U.S. ambassador to the U.N. is found in Lewis L. Strauss, *Men and Decisions,* p. 320.

99–100 James Bryant Conant's views that the nation's energy future was with solar technology and not nuclear power were expressed in a speech he delivered to the American Chemical Society in September 1951. Conant's speech and the views of other leading academicians are contained in hearing transcripts entitled "Atomic Power and Private Enterprise," Joint Committee on Atomic Energy, February 1952, pp. 310–313. See also Lewis Strauss, *Men and Decisions,* p. 320.

101 The American Federation of Labor report warning about the danger of privatization of nuclear technology is from "Report of the Executive Council," 71st Convention, New York, September 15, 1952.

102 The quote on the interests of General Electric and Westinghouse officials to keep "the power consumption rising" is from the Joint Committee on Atomic Energy's 1952 annual report. See "Atomic Power and Private Enterprise," *Bulletin of Atomic Scientists,* May 1953, p. 139.

103 President Eisenhower's attitude toward the TVA and public power can be found in Clyde T. Ellis, *A Giant Step* (New York: Random House, 1966), pp. 107–108.

105 Statements opposing violation of the original provisions of the Atomic Energy Act of 1946 are contained in testimony before the Joint Committee on Atomic Energy during the summer of 1953 and are summarized in *Bulletin of Atomic Scientists,* November 1953, pp. 343–344.

106 The estimate of a $200 billion total subsidy for nuclear power is from Amory Lovins.

The report eventually written on the impact of changing the Atomic Energy Act of 1946 to privatize nuclear power is "Peaceful Uses of Atomic Energy" (U.S. Congress Joint Committee on Atomic Energy), January 1956.

107 The statements by James D. Stietenroth were made in testimony given before the Senate Judiciary Committee and can be found in

PAGE

"Monopoly in the Power Industry. U.S. Senate Interim Report of the Subcommittee of the Committee of the Judiciary on Anti-trust and Monopoly. An Investigation into Monopoly in the Power Industry" (Washington, D.C.: U.S. Government Printing Office, 1955), pp. 11–15.

108 The battling over transmission lines and suspicions about illegal dealings of the Secretary of the Interior are found in "Certain Activities Regarding Power Departments of the Interior," U.S. House Committee on Government Operations, March 28, 1956, and also in "Efforts to Influence the Secretary of the Interior," July 16–26, 1956.

111 A discussion on proposals to weaken the Public Utility Holding Company Act of 1935 is contained in "Proposed Amendments to the Public Utility Holding Company Act of 1935; Hearings before Subcommittee of Interstate and Foreign Commerce Committee," U.S. Senate, April 17–24, May 18, and May 24, 1956.

113 The statement by Lewis Strauss that the opposition to the Fermi breeder reactor was "the first [indication] that private development of atomic power would be fought" is found in Lewis Strauss, *Men and Decisions,* p. 324.

115 Lewis Strauss held out the threat that the Atomic Energy Commission would build reactors if private industry did not move ahead; see "Surprise from Strauss," *Newsweek,* December 24, 1956, p. 19.

For additional information on the development of nuclear power in the postwar era see Anthony Cave Brown and Charles B. McDonald, eds., *Secret History of the Atomic Bomb* (New York: Dial Press/James Wade, 1977); Arthur H. Compton, *Atomic Quest* (Oxford, England: Oxford University Press, 1956); Leslie R. Groves, *Now It Can Be Told: The Story of the Manhattan Project* (New York: Harper, 1962); Richard Hewlett and Oscar Anderson, Jr., *History of the United States Atomic Energy Commission,* 1939–1946 (University Park: Pennsylvania State University Press, 1962); Alice Kimball Smith, *Peril and Hope: The Scientists' Movement in America: 1945–47* (University of Chicago Press, 1965); and *U.S. Atomic Energy Commission: In the Matter of J. Robert Oppenheimer* (Washington, D.C.: U.S. Government Printing Office, 1954).

Chapter 5 Limits to Growth

PAGE

118 Barry Commoner's quote is from *Science and Survival* (New York: Viking, 1966), p. 3.

119 President Kennedy's statement on the need to triple the nation's power output by 1980 is from *Public Papers of the Presidents of the United States—John F. Kennedy—1961* (Washington, D.C.: Office of

PAGE

the Federal Register, National Archives, and Records Service, GSA, 1962), p. 118.

Difficulties in the "big reactor program" were noted in "Seaborg's AEC: Atoms for War or Peace," *Newsweek,* October 16, 1961, p. 65.

120 Warnings sounded on the need for public power systems to join together for self-preservation are found in "Tie or Die, Systems Advised," *Public Power,* July 1960, pp. 13–14.

121 The new calls for a national power grid were noted in *Electrical World,* May 8, 1961, p. 47.

Edwin Vennard's equation of public power systems with the first step to "socialism" is found in Senator Lee Metcalf and Vic Reinemer, *Overcharge* (New York: David McKay Company, 1967), pp. 102, 110.

122 For a full discussion of the Hanford, Washington, generating plant fight see Bonnie Baack Pendergass, op. cit.; see also the 1961–1963 issues of *Public Power.*

Information on the price-fixing scandal in the electric equipment manufacturing industry can be found in John Herling, *The Great Price Conspiracy: The Story of the Anti-Trust Violations in the Electrical Industry* (Washington, D.C.: Robert B. Luce, Inc., 1962). See also John Grant Fuller, *The Gentlemen Conspirators: The Story of Price Fixers in the Electrical Industry* (New York: Grove Press, 1962).

122–123 Kennedy's statement on the progress of electrification and the importance of electricity can be found in *Public Papers of the Presidents of the United States—John F. Kennedy—1962,* p. 623.

123–124 Background information on public opposition that the nuclear industry faced over plants for New York City and Bodega Bay, California, can be found in contemporary issues of *Forum Memo to Members,* published by the Atomic Industrial Forum. These particular quotes are from the July 1963 issue, pp. 7–10.

125 For a full account of the private power companies' efforts to undermine public power in the 1960s see Metcalf and Reinemer, op. cit., pp. 92–226. See also the introduction to the 1964 edition of Ernest Gruening's *The Public Pays* (New York: Vanguard Press, 1964).

126 Fred G. Clark's statement and the number of power companies providing funds to his program are found in Metcalf and Reinemer, op. cit., p. 176.

126–127 A summary of the Massachusetts Institute of Technology Center for Policy Alternatives study is contained in *Future Utility Requirements* (Stamford, Conn.: Business Communications Co., 1974), pp. 111–112.

127 Kinsey Roberts's efforts to rally financial support for right-wing political candidates is found in *Electrical World,* October 24, 1963, p. 14.

PAGE

128 Senator Barry Goldwater's encouragement for private power companies to form a political network and his views and information on the film *The Welfare State* are contained in Metcalf and Reinemer, op. cit., pp. 163, 225. See also *Electrical World,* October 14, 1963, p. 14.

130 President Lyndon Johnson's note on his discussion with Glenn Seaborg and the acceleration of the reactor program is found in *Public Papers of the Presidents—Lyndon Johnson—1963–64,* GSA, 1965, p. 671.

131 The brief note on the accident at Fermi can be found in "Radiation Surge Stops Reactor," *Detroit Free Press,* October 7, 1966, p. 3A. For a complete account of the accident see John C. Fuller, *We Almost Lost Detroit* (New York: Reader's Digest Press, 1975).

131–132 Information on the Northeast blackout of 1965 is contained in "Northeast Power Failure: A Report to the President," Federal Power Commission (Washington, D.C.: U.S. Government Printing Office, 1965). Information on other power failures during that period can be found in additional reports of the Federal Power Commission, including "Power Interruption: Pennsylvania—New Jersey—Maryland Interconnection, June 5, 1967."

133 For an analysis of the manipulation of power pooling arrangements see Mark Messing, H. Paul Friesma, and David Morell, *Centralized Power* (Cambridge, Mass.: Olegeschlager, Gunn and Hain, 1972).

134 The example of Westfield, Massachusetts, is from testimony on the Utility Consumers' Counsel Act of 1969, held before a subcommittee of the U.S. Senate Committee on Government Operations, 91st Congress, 1st session, S. 607, pp. 36–49.

Further information on the new mergers and holding company formation in the late 1960s can be found in "Utilities Enter New Merger Era," *Electric Power and Light,* September 1968, pp. 120–129.

136 For a full account of Drs. John Gofman and Arthur Tamplin see *Poisoned Power: The Case Against Nuclear Power Plants Before and After Three Mile Island* (Emmaus, Pa.: Rodale Press, 1979). See also Anna Gyorgy, *No Nukes* (Boston: South End Press, 1978).

139–140 The summary proceedings of the Sierra Club conference in Vermont in January 1971 were published as *Towards an Energy Policy* (San Francisco: Sierra Club, 1973).

140–141 President Nixon's statement on the restructuring of the industry in fifteen to twenty super generating and transmission companies is found in the 50th anniversary issue of *Electric Light and Power,* September 1972, p. 65. The statement of Secretary of Agriculture Earl

PAGE

Butz is from the same source, p. 72, and the additional federal official's comment on the involvement of power companies in local government and planning can be found on p. 90.

Chapter 6 Citizen Rebellion

PAGES

144–145 The firsthand account of the April 1977 Clamshell demonstration is based on the authors' recollections and eyewitness reports published in the *Boston Globe* and *Manchester Union Leader,* April 29–May 15, 1977.

146–147 Information on the early history of the antinuclear movement is drawn from Anna Gyorgy and Friends, *No Nukes: Everyone's Guide to Nuclear Power* (Boston: South End Press, 1979); Richard S. Lewis, *Nuclear Power Rebellion: Citizens vs. Atomic Industrial Establishment* (New York: Viking Press, 1972); and Jerome Price, *The Anti-Nuclear Movement* (Boston: Twayne Publishers, 1982).

The description of the early history of the New England Coalition on Nuclear Pollution and its role as intervenor in the Vermont Yankee Licensing hearings is based on an interview with Diana Sidebotham, a founding member and former president of NECNP, December 8, 1984.

147 Ralph Nader's statement at the first Critical Mass conference is found in *Critical Mass—The Citizen Movement to Stop Nuclear Power—Newsletter,* vol. 1, April 1975.

148 A full account of "Incident at Browns Ferry: Alabama's Nightmare in Candlelight," written by David D. Comey, is found in *Not Man Apart,* September 1975.

148–149 Quotes from GE defectors are found in *Testimony of Dale G. Bridenbaugh, Richard B. Hubbard, and Gregory C. Minor Before the Joint Committee on Atomic Energy, February 18, 1976, Washington, D.C.* (Reprinted by Union of Concerned Scientists, Cambridge, Mass.); *Testimony of Robert D. Pollard Before the Joint Committee on Atomic Energy, February 18, 1976,* has also been reprinted by the Union of Concerned Scientists.

149 A list of corporate contributors against California's Proposition 15 is found in "Nuclear Power: Who Really Won in California," *Powerline,* vol. 2, June–July 1976, pp. 2–3, and Robert Walters, "Taking the Initiative on Nuclear Power," *National Journal,* June 12, 1976, p. 827. For information on federal efforts to influence the outcome of the referendum see *Report to the Senate Committee on Energy and Natural Resources by the Comptroller General of the United States: Federal Attempts to Influence the Outcome of the June 1976 California Nuclear Referendum, January 27, 1978.*

PAGE

149–150 Harvey Wasserman's statements concerning mass civil disobedience are found in his article "The Clamshell Alliance: Getting It Together," *The Progressive,* vol. 41, September 1977, pp. 14–18.

Source materials for the Seabrook hearings and the founding of the Clamshell Alliance include Gyorgy, op. cit., pp. 395–400; Wasserman, op. cit., pp. 14–18; and Stephen Zunes, "Seabrook: A Turning Point," *The Progressive,* vol. 42, September 1978, pp. 28–31. The authors also benefited from interviews and discussions with founding members of the Clamshell Alliance, including Guy Chichester, Rennie Cushing, Jr., Paul Gunther, Sam Lovejoy, Anna Gyorgy, and Bob Backus, the attorney for the intervenors in the Seabrook construction permit hearings, December–January 1984–5.

151 Information on the international antinuclear movement is found in Gyorgy, op. cit., pp. 297–378, and Harvey Wasserman, *Energy War: Reports from the Front* (Westport: Lawrence Hill & Company, 1979), pp. 151–184.

152–153 For the role of labor unions in the antinuclear movement see Steve Askin, "Labor's Disenchantment with Nuclear Power," *The Progressive,* vol. 45, March 1981, pp. 43–45, and "Solidarity Forever—And No Nukes!" *Powerline,* vol. 6, November 1980, pp. 5–7. The economic impact and the impact on employment of nuclear power, conservation, and other energy options are examined in Steven Buchsbaum and James W. Benson, *Jobs and Energy* (New York: Council on Economic Priorities, 1979).

154–156 Source materials for the controversy at Big Mountain, Arizona, include Pelican Lee, "Navajos Resist Forced Relocation: Big Mountain and Joint Use Area Communities Fight Removal" (Big Mountain Support Group, Albuquerque, N.M., n.d.); Richard Brady, "Hopi Resolution Scuttles Fence Plan," *Arizona Daily Sun,* August 15, 1984; "Fencing Fiasco Flares Again," *Big Mountain News* (Fall 1984); and Catherine Elston, " 'Take Us Away from the Land, We Die,' " *Navajo-Hopi Observer,* August 1, 1984; See also *Navajo Relocation Review,* a special report by the Navajo-Hopi Task Force, the Land Dispute Commission, and the *Navajo Times,* July 1982. The authors also benefited from an interview with Chris Shuey, research associate at the Southwest Research and Information Center, January 23, 1985.

156 For information on the corporate energy exploitation of the Black Hills see Michael Garitty, "Stopping the Sacrifice of the Black Hills," *Win Magazine,* vol. 16, July 1, 1980, pp. 17–19. Information on the present controversy over storage of nuclear waste in South Dakota is drawn from an interview with Karen Funk, National Congress of American Indians, Washington, D.C., January 1985.

156–157 Data on the adverse physical effects of extra-high-voltage transmission is found in Louise B. Young and H. Peyton Young, "Pollution by

PAGE

Electric Transmission," *Bulletin of Atomic Scientists,* December 1974; Gladwin Hill, "Shocker—Effects of Power Lines on People Pondered," *New York Times,* November 10, 1975; and Susan Schiefelbein, "The Invisible Threat: The Stifled Story of Electric Waves," *Saturday Review,* September 15, 1979. See also *Electrical Environment Outside the Right of Way of CU-TR-I, Report 5, May 1985* (Power Plant Siting Program, Minnesota Environmental Quality Board, Saint Paul, Minnesota); *Status Report,* New York State Department of Health Research Program, to determine public health risks to humans from exposure to electric and magnetic fields, April 1, 1985.

157–158 Source material for power-line protests includes Louise B. Young, *Power Over People* (New York: Oxford University Press, 1973); "The Public and the Powerline: A Classic Case of Government against the People . . . ," *Mother Earth News* (March–April 1980), pp. 68–69; Richard Severo, "Powerline Project Has Many North Country People Boiling," *New York Times,* January 20, 1977; "Hundreds Protest Along PASNY Power Line Route," *Rochester Patriot,* September 14–27, 1977; "Citizens Battling High Voltage Power Line," *Washington Resource Report,* July 1978, p. 8; and "Marcy-South Article VII Update," *Citizen Power of Orange County Newsletter,* December 21, 1983.

158–160 For the Minnesota fight see Barry M. Casper and Paul Wellstone, *Powerline: The First Battle of America's Energy War* (Amherst: University of Massachusetts Press, 1981), and Harvey Wasserman, "Revolt of the Bolt Weevils," *Rolling Stone* (August 9, 1979), pp. 38–40; see also back issues of *Hold That Line* (power-line protest newsletter of Central Minnesota). For New England see "Power Lines in Vermont?" (pamphlet published by Northeast Kingdom Energy Alliance, Greensboro Bend, Vt., n.d.) and Jean Lowell, "Rural Vermont and the Corridors of Power," *Direct Current,* Summer 1981, pp. 1, 3. The authors also benefited from discussions with Alice Tripp and George Crocker of the General Assembly to Stop the Powerline; Doris Delaney, chairperson of Citizens' Power of Orange County, Court Richardson, Vermont PIRG, January 1985; and Hal Lynde, a member of the Powerline Awareness Campaign.

161–163 The information on utility industry surveillance is based on an interview with Daniel Sheehan, July 1984, and Paul Altmeyer's report "Utility Security," *NBC News,* December 25, 1977. See also Vasil Pappas, "Utilities Generate Sparks by Keeping Close Eye on Critics," *Wall Street Journal,* January 11, 1979, and Frances Cerra, "Two Million Dollar Damage Suit by LILCO Challenges Opponents of Shoreham A-Plant," *New York Times,* October 9, 1981.

164 Nuclear Regulatory Chairman Joseph Hendrie's statement concerning TMI is found in *President's Commission on the Accident at Three Mile Island, Report* (Washington, D.C.: U.S. Government Printing Office, 1979), p. 337.

PAGE

164–165 Source material for the accident at TMI and other setbacks to the nuclear industry includes Mark Hertsgaard, *Nuclear Inc., The Men and Money Behind Nuclear Energy* (New York: Pantheon Books, 1983); Daniel Ford, *Three Mile Island, Thirty Minutes to Meltdown* (New York: Penguin Books, 1981); Harvey Wasserman, "The Anti-Nuclear Movement Approaches Critical Mass," *New Age,* vol. 4, June 1979, pp. 22–32, 52–57; "Three Mile Island: Is It the Downfall of Nuclear Power?" *Critical Mass Journal,* vol. 5, special issue, no. 5; Alden Meyer, "Lemon Tree Yields Costly Harvest," *Powerline,* vol. 4, May 1979, pp. 5–6; and "Lightening the Nuclear Sled: Some Uses and Misuses of the Three Mile Island Accident," remarks by Commissioner Peter A. Bradford, U.S. Nuclear Regulatory Commission, before the Seminar on Problems of Energy Policy, New York University, New York, November 21, 1979.

166 The National Council of Churches' statement on energy policy is found in "The Ethical Implications of Energy Production and Use" (Governing Board of the National Council of Churches in U.S.A., May 11, 1979).

168 The quote from Nunzio Palladino concerning TMI is found in Kathy Chamberlain, "Conference Fuels Debate over TMI Impact," *Critical Mass Journal,* vol. 5, September 1979, p. 5.

168–171 For information on the nuclear industry's massive advertising campaign and the Safe Energy Communication Council's efforts to counter this propaganda see Hertsgaard, op. cit., pp. 186–205; David Burnham, "Industry and Government Efforts for Nuclear Power Draw Scrutiny," *New York Times,* May 23, 1984; Arlen J. Large, "Nuclear Power Plant Advocates and Foes Start Battling through TV Commercials," *Wall Street Journal,* December 9, 1982; William J. Lanovette, "Industry Goliath, Environmental David Girding for Battle over Nuclear Power," *National Journal,* May 9, 1983; Scott Ridley, "Industry Media Explosion Set for July," *Powerline,* vol. 8, May 1983, pp. 3, 12; Saundra Gail Patton, "CEA Called before House Subcommittee," *Powerline,* vol. 9, August/September 1983, p. 3; and Alan Nogee, "EEI Prepares Ad Onslaught," *Powerline,* vol. 9, December 1983/January 1984, pp. 1, 4.

172 A description of the Winner-Wagner PR firm's efforts to defeat citizen initiatives is found in "Cub, Nuclear Waste Initiatives on State Ballots," *Powerline,* vol. 10, September/October 1984, p. 10, and "Utility Campaign Defeats Missouri Initiative," *Powerline,* vol. 10, November/December 1984, p. 8.

173 The overview of the public power fight in Massachusetts is based on the authors' recollections and discussions with Connie Whitney and other members of the Municipal Power Coalition.

PAGE

173–174 Source material for other organizing efforts to reform public power includes Jeff Brummer, "Oklahoma Coop Members Sweep Recall," *Powerline,* vol. 8, December 1982/January 1983, pp. 5, 13; Brummer, "Chocktaw Syndrome, Insurgent Members Shock Co-op Board," *Powerline,* vol. 8, June 1983, pp. 1, 9; "Blacks Battle for Control of Delta Electric Co-op," *Rural America,* September/October 1983, p. 4.; Tim Kalich, "Delta Electric Board to Hold New Elections for All 11 Seats," *Greenwood Commonwealth,* May 28, 1983; Tony Tharp, "Blacks Campaign for Election on Promise of Lower Utilities," *Clarion Ledger,* March 29, 1983; and *Ratewatcher,* vol. 2, October, November, December issues—published by Coalition for Fair Utility Rates.

175 Dick Munson's statement on independent power producers is found in Michael Phillips, "Small Is Sensible: Richard Munson on the New Era of Electricity Generation," *RAIN,* Spring 1986, p. 35.

176 Lando Zech, Jr.'s, remarks on the need for 200 new coal or nuclear plants by the year 2000 are found in "Remarks by Commissioner Lando W. Zech, Jr.," Edison Electric Institute's Governmental Affairs Conference, Marco Island, Fla., January 1986.

176–177 Comments of Amory Lovins and Scott Sklar concerning the transition to solar and conservation strategies are from interviews with one of the authors, June–July, 1984. See also Amory Lovins, "Least Cost Electrical Services as an Alternative to the Braidwood Project," testimony before the Illinois Commerce Commission, July 3, 1985. (Old Snowmass, Colo.: Rocky Mountain Institute).

Chapter 7 Regulation: The Impossible Task

PAGES

181–182 The comments of Mary Metcalf are from an interview with one of the authors, July 5, 1984.

182–184 Information on the Seabrook bailout plan and the appointment of John Nassikas to the New Hampshire Public Utility Commission is drawn from an interview with Kirk Stone, Director of the New Hampshire Campaign for Ratepayer Rights, September 1984, and Scott Ridley, "Seabrook Bailout Plan Falters," *Powerline,* vol. 10, September/October 1984, pp. 1, 8, 9.

184 The comments of Michael Johnson, Pennsylvania Public Utility Commissioner, are from an interview with one of the authors, August 15, 1984; see also Bart Sullivan, "Maverick PUC Member Fights Hikes," *Sentinel Weekender* (Carlisle, Pa.), September 24, 1983, and "Penn. Activists Honor PUC Commissioner," *Powerline,* vol. 8, December/January 1983, p. 16.

185 Robert Eye's statement concerning rate shock and the Wolf Creek plant is found in Alan Nogee, "Kansas Regulators Bite Wolf," *Powerline,* vol. 11, September/October 1985, pp. 1, 3.

PAGE

185–186 Peter Navarro's statements concerning rate suppression and the Texas PUC are found in his "Utility Bills: The Real Price of Electricity," *Wall Street Journal,* January 13, 1983; "The Coming Electric Crisis," ibid., October 17, 1984. Details on Montana's PUC decision concerning the Colstrip 3 plant are found in "Montana Tosses Colstrip Three Out of Rate Base," *Powerline,* vol. 10, September/October 1984, p. 3.

187 Data on the jurisdiction and scope of utility regulation and size of staff is found in Geneva Bierlein, ed., *1982 Annual Report on Utility and Carrier Regulation* (Washington, D.C.: National Association of Regulatory Utility Commissioners, 1982). The Maine comparison of dollars spent in relation to utility revenues is from *Report to the Joint Standing Committee of Public Utilities, Maine Legislature: Public Utilities Commission Annual Report, March 1984.*

188–189 The comments by Harvey Salgo, attorney, Greg Palast, public utility analyst, and Charles Komanoff, energy economist, are from interviews with the authors, April 26, July 6, and August 10, 1984.

189–190 The information on regulatory scams is drawn from interviews with Greg Palast, August 10, 1984, and Peter Bradford, chairman of the Maine Public Utility Commission, July 20, 1984; see also "Maine PUC Investigates Central Maine Power," *Powerline,* vol. 8, December 1982/January 1983, p. 16; "Maine Utility Scandal Ends with Fine, Resignation," ibid., vol. 9, October/November 1983, p. 12; "N. Indiana Public Service Co. Called on Strike Profiteering," ibid., vol. 7, June 1982, p. 12; and "Findings of Fact: Conclusions of Law and Order in the Matter of Public Service Company of New Mexico's Purchase of a Lear Jet 35-A before the New Mexico Public Service Commission," case no. 1468, March 29, 1978.

191 Source materials on phantom taxes include Environmental Action Foundation, *Elected Utility Phantom Taxes, 1982* (April 1984), and Jerry Knight, " 'Phantom Taxes': Utilities' Con Game," *Washington Post,* May 9, 1984.

For a list of contributors to the pronuclear Committee for Energy Awareness PR campaign for calendar year 1983 see *Powerline,* vol. 9, September 1983, p. 3.

194 The quote from Representative Emanuel Celler concerning the failure of the FPC to regulate water power development in the 1920s is found in Bernard Schwartz, *Economic Regulation of Business and Industry,* vol. III (New York: Chelsea House Publishing, 1973), p. 2,051.

195 For details of President Franklin Roosevelt's efforts to strengthen state and federal regulations see James C. Bonbright, *Public Utilities and the National Power Policies;* The 20th Century Fund, *Electric Power and Government Policy* (New York: 20th Century Fund, 1948); Rob-

PAGE

ert D. Baum, *The FDC and State Utility Regulation* (Washington, D.C.: American Council on Public Affairs, 1942); and Harold H. Young, *Forty Years of Public Utility Finance* (Charlottesville, Va.: University of Virginia Press, 1965).

197 Information on Wisconsin and other state efforts to implement marginal cost pricing is found in Charles J. Cicchetti, William J. Gillen, and Paul Smolensky, *The Marginal Cost and Pricing of Electricity: An Applied Approach* (Cambridge, Ballinger Company, 1977), and Werner Sichel, *Public Utility Ratemaking in an Energy Conscious Environment* (Boulder, Colo.: Westfield Press, 1979).

A description of the changes in the economic, social and political environment within which public utility regulation took place during the 1970s is found in Charles F. Phillips, Jr., "The Changing Environment of Public Utility Regulation: An Overview," in Albert L. Danielsen and David R. Kamerschen, eds., *Current Issues in Public Utility Economics* (Lexington, Mass.: Lexington Books, 1983), pp. 25–39.

197–198 Statistics on the oil embargo and its effects on Consolidated Edison and other utilities are found in Douglas D. Anderson, "State Regulation of Electric Utilities," in James Q. Wilson, ed., *Politics of Regulation* (New York: Basic Books, 1980), pp. 22–23, and Leonard S. Hyman, *America's Electric Utilities: Past, Present and Future* (Arlington, Va.: Public Utilities Report), pp. 113–115.

198 Data on fuel adjustment increases is found in *Electric and Gas Utility Rate and Fuel Adjustment Clause Increases, 1977 Committee Report,* Senate, 95th Congress, 2nd session, Subcommittee on Intergovernmental Relations and Subcommittee on Energy, Nuclear Proliferation and Federal Services (Washington, D.C.: U.S. Government Printing Office, 1978).

198–199 The source material for consumer utility reform campaigns includes Richard E. Morgan, *The Rate Watcher's Guide: How to Shape Up Your Utility's Rate Structure* (Washington, D.C.: Environmental Action Foundation, 1980); Donald Anderson, *Regulatory Politics and Electrical Utilities: A Case Study in Political Economy* (Boston: Auburn House Publishing Company, 1981); *Powerline,* vols. 1–10, July 1975 to the present; and William T. Gormley, Jr., *The Politics of Public Utility Regulation* (University of Pittsburgh Press, 1983), pp. 37–64.

199 The testimony of Maine PUC chairman Peter Bradford and other critics of the Utility Reorganization Act of 1975 is found in *The Utilities Act of 1975/Senate Hearings.* Hearings before Subcommittee on Intergovernmental Relations and the Subcommittee on Reports, Accounting and Management of the Committee of Government Operations. U.S. Senate, 94th Congress, 1st session (Washington, D.C.: U.S. Government Printing Office, 1975).

PAGE

200–201 Source materials for legal battles fought over implementing PURPA include Ed Thompson, "Is Utility Reform Unconstitutional?" *Powerline,* vol. 6, April 1981, pp. 1, 6; Richard Morgan, "Court Strikes PURPA Rules for Co-generation," ibid., vol. 7, March 1982; Richard Morgan, "Landmark Law Survives Mississippi States-Rights Challenge," ibid., vol. 7, July 1982, pp. 1, 5. See also Jay B. Kennedy, "DOE Mandates: Taking Power Away from the States," *Public Utilities Fortnightly,* November 6, 1980, pp. 11–14.

201–202 Information on industry ties of President Reagan's administrative appointees is found in Ronald Brownstein and Nina Eaton, *Reagan's Ruling Class* (Washington, D.C.: Presidential Accountability Group, Ralph Nader, 1982); "Some of the President's Men," *Powerline,* vol. 6, February 1981, pp. 5, 7; "FERCed," ibid., vol. 7, August 1981, p. 5.

202 For a critique of Reagan's energy policies see Friends of the Earth, et al., *Indictment: The Case Against the Reagan Environmental Record* (Washington, D.C., 1981).

James Edwards's remarks (former Secretary, U.S. Department of Energy) are found in Brownstein and Eaton, op. cit., p. 145.

202–203 Julian Greenspun's statement concerning the prosecution of NRC criminal cases is found in Howard Kurtz, "NRC Officials Avoid Pursuit of Wrongdoing, Critics Say," *Washington Post,* April 8, 1986. Commissioner James K. Asselstine's remark is found in James K. Asselstine, "Regulating Nuclear Industry with a Reaction of Dissent," *Washington Post,* June 9, 1986.

203 Representative Ed Markey's remarks are found in Michael Weisskopf, "U.S. Reactors' Record Called Worst Since 1979," *Washington Post,* May 4, 1986.

204 Michael Pertschuk's statements on regulation are found in Molly Sinclair, "Quotations from Chairman Pertschuk," *Washington Post Magazine,* October 28, 1984, p. 12.

205 Information on the "gang of four" and President Reagan's utility bailout plan is drawn from Alan J. Nogee, "Sweeping Reforms Would Limit State Authority: Reagan Administration to Unveil Radical Utility Bailout Plan," *Powerline,* vol. 8, October 1982, pp. 1, 6–7; "Doe Claims Broad Support for Principles—Reagan Working Group Focuses Bailout Plan," ibid., vol. 8, December 1982/January 1983, pp. 1, 12; and "Working Group Update: Bailout Bogs Down," ibid., vol. 8, March 1983, pp. 1, 7.

205–206 The quotes from David Grubb and Governor Jay Rockefeller and the description of the Consumer Action Group's effort to reform utility rates in West Virginia are from Margaret Peterson, "Activist David Grubb," *The Congress Watcher,* July/August 1983, p. 13, and Scott

PAGE

Ridley, "Almost Heaven—W. Virginia Passes Unprecedented Reform," *Powerline,* vol. 8, May 1983, pp. 1, 9.

206 The information on elected utility commissions is drawn from *Illinois Public Action Council, The Record on Elected and Appointed Utility Commissioners, A Public Action Report, February, 1981,* and Jeff Brummer, "No Regulation Without Representation!: From Maine to Oregon, Citizens Demand Elected Utility Commissions," *Powerline,* vol. 7, June 1981, pp. 1–11. For a model plan for establishing elected commissions see Lee Webb, "A Model Agenda for State and Local Governments—A State Public Utility Commission Act," *Ways and Means,* November–December 1981, pp. 3–4, and "Court Ruling May Derail Utility Panel," *Portland, Maine Press Herald,* March 1, 1986.

207 For how methods of selection of utility regulatory commissioners may determine the regulatory climate in the states see Joseph C. Samprone, Jr., and Nancy Riddell-Dudra, "State Regulatory Climate: Can It Be Predicted?" *Public Utilities Fortnightly,* October 8, 1981, pp. 41–42.

206–207 Information on Citizen Utility Boards Is Drawn from "Cub Takes Its First Steps," *Powerline,* vol. 7, August 1981, pp. 3, 7; Scott Ridley, "Cub Making Tracks in 9 States—Citizens Utility Board Picks Up Momentum and Critics," ibid., vol. 7, April 1982, pp. 1, 7; "New Cub-like Group in Southern California," ibid., vol. 8, May 1983, p. 12; and "Illinois to Have a Cub," ibid., vol. 9, August/September 1983, p. 12. The authors also benefited from an interview with Peter Anderson, President, Wisconsin's Environmental Decade, Madison, Wisconsin, May 26, 1984.

208 The quote of Rogers C. B. Morton to "service" and not "regulate" industry is from David S. Freeman, *Energy: The New Era* (New York: Walker, 1974), p. 190.

Information on the National Association of Regulatory Utility Commissioners Investigation is found in "NARUC Probe Heightens Controversy: Four States Slash Edison Electric Dues," *Powerline,* vol. 9, June/July/August 1984, pp. 1, 4; "Showdown Looms on EEI Probe," ibid., vol. 10, March/April 1985, p. 10; "Controversy Clouds First EEI Audit," ibid., vol. 10, July/August 1985, p. 8; and "Lobbying Called EEI's Raison d'Être," ibid., vol. 10, November/December 1985, p. 7.

In addition to the above citations and works cited in the notes to Chapter 2, the following general accounts were useful in tracing the origins and development of public utility regulation: Irston R. Barnes, *The Economics of Utility Regulation* (New York: F. S. Crofts & Company, 1942); James C. Bonbright and Gardiner C. Means, *The Holding Company: Its Public Significance and Its Regulation* (New York:

PAGE

McGraw-Hill, 1932); Martin T. Farris and Roy J. Sampson, *Public Utilities: Regulation, Management, and Ownership* (Boston: Houghton Mifflin Company, 1973); Lloyd Wilson, James Herring, and Roland B. Eutsler, *Public Utility Regulation* (New York: McGraw-Hill, 1938); and I. Leo Sharfman, "Commission Regulation of Public Utilities: A Survey of Legislation," *The Annals*, vol. 53, May 1914.

For the 1950s and 1960s see Metcalf and Reinemer, Harry M. Trebing, ed., *Essays on Public Utility Pricing and Regulation* (East Lansing: Michigan State University, 1971), and Warren J. Samuels and Harry M. Trebing, eds., *A Critique of Administrative Regulation of Public Utilities* (East Lansing, Mich.: MSU, Public Utilities Papers, 1972).

Among the many works on current regulatory issues, such as deregulation, diversification, and regional regulation, the following were most useful: "The Electric Utility Industry: Rethinking Regulation," *Lehman Brothers Kuhn Loeb Research—Industry Review*, February 1983; *Deregulation in the Electric Power Industry*, conference proceedings, Bethesda, Md., November 16–17, 1982 (sponsored by Mitre Corporation and Edison Electric Institute); Llewellyn King, "Beyond Ratemaking," *Energy Daily*, April 13, 1982, p. 2; National Governors' Association Committee on Energy and Environment, *An Analysis of Options for Structural Reform in Electrical Utility Regulation, Report of the NGA Task Force on Electric Utility Regulation, January 1983;* Michael B. Meyer, "A Modest Proposal for the Partial Deregulation of Electric Utilities," *Public Utilities Fortnightly*, April 14, 1983, pp. 3–26; Stanley York and J. Robert Malko, "Utility Diversification: A Regulatory Perspective," *Public Utilities Fortnightly*, January 6, 1983, pp. 15–19; and Larry J. Wallace, "Re-Regulation of the Electric Utility Industry: A Neglected Alternative," ibid., November 25, 1982.

Additional works on the history of the Federal Power Commission that were useful include Gifford Pinchot, "The Long Struggle for Effective Federal Water Power Legislation," *George Washington Law Review*, vol. 14, December 1945; Burton K. Wheeler, "The FPC as an Agency of Congress," ibid.; and Stephen G. Reyer and Paul W. Macavoy, *Energy Regulation by the Federal Power Commission* (Washington, D.C.: Brookings Institute, 1974).

Chapter 8 Wall Street: The Stall of the Dividend Machines

PAGE

210 The statement by Eugene Meyer is from "Will Investors Ever Put Money in Utilities Again?" *Energy Daily*, August 20, 1984, pp. 2–3.

211 For general background on the financing of the utility industry from Wall Street's perspective see Leonard Hyman, *America's Electric*

PAGE

Utilities: Past, Present, Future (Arlington, Va.: Public Utilities Reports, 1983).

212 The speech given by Frank Vanderlip on the need for capital to finance utility expansion in 1913 is in booklet form, Frank A. Vanderlip, *Financing Electricity* (New York: The National City Bank of New York, September 5, 1913).

214–215 The interrogation of Hugh Magill of the American Federation of Utility Investors is taken from "Utility Lobby Investigation," U.S. House Committee on Rules, July 25, 1935 (Washington, D.C.: U.S. Government Printing Office).

215 The investigation that uncovered continuing ties between the financial community and the power industry in 1940 is found in "The Problem of Maintaining Arm's Length Bargaining and Competitive Conditions in the Sale and Distribution of Registered Public Utility Holding Companies and Their Subsidiaries," Securities and Exchange Commission, Public Utilities Division, 1940. See especially p. 12 and Appendix F-1.

216 Information on the 1955 Dixon-Yates controversy can be found in "Monopoly in the Power Industry," interim report of the Subcommittee of the Judiciary on Anti-Trust and Monopoly, 83rd Congress, 2nd session, 1955, pp. 87–93.

217 Congressman Michael Harrington's statements can be found in testimony before the U.S. Senate Committee on Government Operations, "Corporate Disclosure Hearings," part 3, August 14, 1974 (Washington, D.C.: U.S. Government Printing Office).

218 The estimate of 1,000 shared directors between bank and power companies is used by many sources, including *Electric Utility Week*, February 13, 1984, p. 3.

The series of corporate disclosure hearings conducted by the U.S. Senate provide access to information rarely available to the public. See Hearings Before the Subcommittee on Budgeting, Management, and Expenditures and the Subcommittee on Intergovernmental Relations of the Committee on Government Operations, "Corporate Disclosure Hearings," parts I, II, III, March 21–August 14 (Washington D.C.: U.S. Government Printing Office, 1974); "Corporate Ownership and Control," November 1975; and "Structure of Corporate Concentration: Institutional Shareholders and Interlocking Directorates Among Major U.S. Corporations," December 1980 (Washington, D.C., U.S. Government Printing Office).

220 The figures for donations by Wall Street firms to the campaign against Initiative 391 are from records of the Washington State Public Disclosure Commission.

PAGE

The description of the Chemical Bank event at the Felt Forum in October 1983 is from personal observation.

221 For background on the Three Mile Island financial meltdown and the bailout of General Public Utilities see Tim Metz and Gene G. Marcial, "Prices Fall for Nuclear, Uranium Stocks, but Four-Day Flurry May Have Peaked," *Wall Street Journal,* April 3, 1979; Ann MacLachian, "Bank of America More Skittish Than Most on Nuclear Loans," *Energy Daily,* May 17, 1979, p. 1; and "Hard Times in Nuclearland," *Groundswell* (Washington, D.C.: Nuclear Information and Resource Service), May 1980, pp. 9–10.

221–222 The information on Eugene Meyer is from interview material plus "Testimony of Eugene W. Meyer before the Senate Energy Committee, July 14, 1983: 'Will Investors Ever Put Money in Utilities Again?,'" *Energy Daily,* August 20, 1984, pp. 2–3; "Kidder Peabody Plans to 'Educate' Washington on Nuclear Problems," *Nucleonics Week,* September 13, 1984, pp. 4–5.

223 The information on campaign contributions is drawn from reports filed with the Federal Election Commission in Washington, D.C.

224–226 The information from Eileen Austen of Drexal Burnham Lambert, Inc., is from an interview with one of the authors. Other interviews for this section were done with Jeffrey Whitehorn of the Dreyfus Corporation, Charles Silberstein of the Finance Guaranty Insurance Company, S. Arlene Barnes of First Boston Research Corporation, Mark Luftig of Salomon Brothers, and a number of other analysts, brokers, and investment bankers.

226 For a full analysis of the WPPSS catastrophe from Wall Street's perspective, see Howard Gluckman, "WPPSS: From Dream to Default," *The Bond Buyer,* 1984. See p. 37 for quote.

228 The process for capitalizing an energy transition without the direct participation of Wall Street is detailed by Amory and Hunter Lovins in "Capitalizing the Country's Energy Transition," *The Neighborhood Works,* Spring 1983, pp. 9–18.

229 John Dorfman's statement on Southern California Edison can be found in his book *The Stock Market Directory* (New York: Doubleday, 1982), p. 411.

230 For background on the General Electric purchase of RCA Corp. and its NBC television and radio networks see "GE Completes RCA Purchase," *Washington Post,* June 10, 1986. Information on the formation of new holding companies is from the Securities and Exchange Commission.

230–231 Analysis of the formation of new holding companies can be found in testimony presented by Scott Ridley. See "U.S. House Subcommittee

PAGE

on Energy Conservation and Power and the Subcommittee on Telecommunications, Consumer Protection, and Finance on Proposed Amendments to the Public Utility Holding Company Act of 1935, October 31, 1983" (Washington, D.C.: U.S. Government Printing Office), pp. 332–350.

233 Information on the new wave of customer-ownership programs to neutralize consumer opposition can be found in *Electric Utility Week,* July 18 and August 22, 1983.

234 For background on James Spang and the American Society of Utility Investors see Tom Torok, "They're Battling to Bolster Utility Investors," *Philadelphia Inquirer,* May 15, 1984; see also Spang's testimony before the U.S. Senate Subcommittee on Energy Regulation, April 12, 1984.

235 Information on the stock ownership of Commonwealth Edison of Chicago is from the Federal Energy Regulatory Commission *Form 1,* 1982.

The provocative statement by Frank W. Griffith can be found in *Public Utilities Fortnightly,* May 13, 1982, p. 101, and Robert Bigwood's statement in *Public Utilities Fortnightly,* June 8, 1983, p. 109.

236 Griffith Morris's statement on the continued need for nuclear plants was delivered to the Sixth National Utility Conference on the Crisis in Electric Power, Washington, D.C., June 4, 1984.

Chapter 9 The Coming Clash

PAGE

238 Dr. Edward Teller's statement can be found in an advertisement published in the *Wall Street Journal,* July 31, 1979.

The information on the Chernobyl accident is drawn from a number of sources. See Celestine Bohlen, "Soviets Cite Delays in Evacuation"; Boyce Rensberger, "Thousands of Soviets May Fall Ill"; and Cass Peterson, "Experts See No Danger in U.S.; all in the *Washington Post,* May 1, 1986.

The *NBC/Wall Street Journal* poll showing 65 percent of the people surveyed opposing further nuclear construction was cited in *USA Today,* May 2–4, 1986, p. 1. A *Washington Post/ABC* poll revealed 78 percent of some 1,506 people interviewed opposed further nuclear construction in the United States: *Washington Post,* May 24, 1986, p. A6.

239 Information on the Department of Energy's time line for a resurgence is found in *Strategic Plan for the Civilian Reactor Development Program,* draft, U.S. Department of Energy, October 25, 1985. Additional information on plans to clear barriers is from "Report of Civilian Nuclear Power Subpanel 3; Institutional Challenges," Department of Energy, January 30, 1986.

PAGE

240 The Stonehenge and Atomic Priesthood proposal came from the study "Communication Measures to Bridge Ten Millennia," conducted by Battelle Memorial Institute under a contract from the Department of Energy. A review of the proposal can be found in T. R. Reid, "Warning Earthlings of Atomic Dumps," *Washington Post,* November 11, 1984.

An overview and local responses to the federal government's plans for nuclear waste storage can be found in David F. Salisbury, "Storing Nuclear Waste" (a five-part series), *Christian Science Monitor,* June 24–28, 1985.

Involvement of the Committee for Energy Awareness in nuclear waste siting can be found in the U.S. CEA 1984 Program Plan and also in "CEA Rushing in Where Angels Fear to Tread," *Clarion-Ledger,* Jackson, Mississippi, January 19, 1984, and "Group Clashes with Environmentalists over Waste Dump," *Spokesman-Review,* Seattle, Washington, March 25, 1984.

The description of protests in Maine and quotes are from Bob Cummings, "Brennan Warms to Yankee Closing," *Portland Press Herald,* February 21, 1986, p. 1.

242 The threat of plant shutdowns can be found in the *North American Electric Reliability Council Annual Report, 1984* (Princeton, N.J.: North American Electric Reliability Council).

242–243 J. Hunter Chile's statement can be found in "Build Now to Avoid Blackouts Tomorrow," *Across the Board,* November 1984.

243 Deputy Secretary of Energy Danny Boggs made the statement on the "lights going out" during the Senate Energy and Commerce Committee hearings, July 23, 1985. See "Utility Executives Reveal Shortages, Senate Listens," *Electrical World,* September 1985, pp. 17–21.

Alan Nogee made the allusion to a "protection racket" during an interview on November 11, 1985.

Steven Ferry's comments can be found in "What Power Shortage?" (Letters section), *New York Times,* April 24, 1984.

244 For a full explanation of least-cost alternatives see Amory Lovins, "Least Cost Electrical Services as an Alternative to the Braidwood Project," testimony presented to the Illinois Commerce Commission, July 3, 1985 (Old Snowmass, Colo.: Rocky Mountain Institute).

For the estimate of 200,000-megawatt potential cogeneration see *New Electric Power Technologies, Problems and Prospects for the 1990s,* Office of Technology Assessment (Washington, D.C.: U.S. Government Printing Office, July 1985), p. 198.

245 Charles Komanoff's statements were made in a conversation November 11, 1985.

246 The Edison Electric Institute targets for legislative reform can be found in *Alternative Models of Electric Power Deregulation,* VI 12,

PAGE

Edison Electric Institute, May 1982. The efforts to streamline the nuclear licensing process are contained in the Nuclear Facility Standardization Act of 1985 proposed by the Department of Energy.

247 The "vociferous few" is a term used by Frank Griffith, president of Iowa Public Service Company and former chairman of Edison Electric Institute. See Chapter 8 notes, to p. 235.

248 For comparative figures on taxes paid by public and private power systems see Jeannie Shenk, "Public Power Contributes More," *Public Power,* May–June 1986, p. 25.

248–250 The information on the Long Island confrontation is from interviews and local reporting by *Newsday* writers. For a history of the Shoreham project see Stuart Diamond, "Shoreham: What Went Wrong," *Newsday,* published as a series of reports November 15–22, 1981. See also Maurice Carroll, "State Takeover of Lilco Urged by Republicans," *New York Times,* January 7, 1986.

250–253 Much of the information on New Orleans was gathered in interviews. See also "Major Blackout Is Triggered by Cold," *Times-Picayune,* January 22, 1985; "NOPSI Sued for $100 Million in Blackout," *Times-Picayune,* January 24, 1985; "Blackout Defended by Utilities," *Times-Picayune,* January 29, 1985; and "Time to Buy Out NOPSI," *Gambit,* January 19, 1985.

253–256 The material on Pacific Gas and Electric was gathered from interviews and "New Move to Municipalize PG&E," *San Francisco Bay Guardian,* April 7, 1982; "PG&E Attempts a Coup Against Public Power in San Francisco," ibid., September 11, 1982; "Guess Who's Behind San Franciscans for a Responsible Energy Policy?" ibid., September 11, 1982; and "The Bay Guardian vs. PG&E—The 16 Years' War," ibid., October 6, 1982.

256 Background on the Susanville, California, takeover proposal can be found in Bill Paul, "Electric Utilities Find Market Forces Taking More Important Role," *Wall Street Journal,* February 26, 1986, p. 1.

257 David Talbot and Richard E. Morgan, *Power and Light: Strategies for the Solar Transition* (New York: Pilgrim Press, 1981).

259 The international status of nuclear power was gathered from reports by the International Atomic Energy Agency as well as numerous press reports. Despite the misgivings of local populations, the IAEA says plans are for European Common Market countries to increase nuclear-supplied electricity from 27 percent in 1985 to 40 percent by 1995.

263 Gifford Pinchot's statement can be found in Gifford Pinchot, *The Power Monopoly: Its Make-up and Its Menace* (Milford, 1928), pp. 1–2.

Index